Lotus

Illustrated DICTIONARY *of* BIO CHEMISTRY

Complied and Edited by:
Henry James

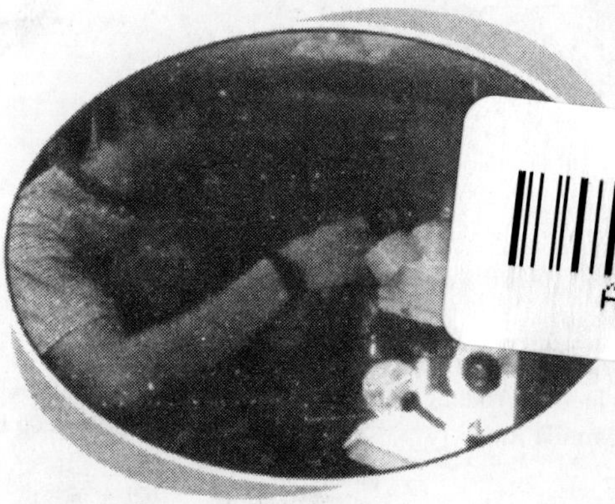

Lotus
PRESS
4735/22, Prakash Deep Building
Ansari Road, Darya Ganj,
New Delhi-110002

Lotus *Illustrated*

DICTIONARY *of*

BIO CHEMISTRY

ISBN 81 89093 15 0

Published by:

Lotus Press : Publishers & Distributors

Unit No. 220, 2nd Floor, 4735/22,
Prakash Deep Building, Ansari Road, Darya Ganj,
New Delhi- 110002, Ph.: 32903912, 23280047, 09811838000
• E-mail : lotus_press@sify.com www.lotuspress.co.in

Printed at : ***Concept Imprint***, Delhi

PREFACE

Biochemistry is the chemistry of living organisms and their vital processes. The process is being carried out in all of us but still we are ignorant about all that. This dictionary of biochemistry is developed in order to apprise the readers with various terms that are associated with biochemistry and which are happening in all of us. This dictionary of biochemistry contains more than 2000 terms closely related with the bio-chemical processes.and associated topics. This is because merely describing the term cannot do justice with that. A conscious effort has been made to develop it in a manner so that it can be easily understood and the thoughts generated matches with the actual processes. To make it so this edition of dictionary comprises exhaustive, well-researched and precise definitions. Besides that, suffixes and prefixes are also given which will certainly help readers in understanding which are new to them or which happen to be not present in the dictionary. The definitions are supplemented by the pictorial representations, which will be an easy tool to visualise the concerned process in a more effective way. The pictorial representation is not just for emphasising the very fact mentioned in the dictionary but it is include imparting extra knowledge.

This dictionary is meticulously developed, keeping in mind the interest and requirement of students, teachers, researches and all the other persons who in a way or other are of biological processes. And this dictionary will certainly help them in their causes, because it is developed with much effort and great care with lot of hard work. In the end we wish all the readers a very happy reading.

G°'

standard free-energy change.

α-helix

a helical conformation of a polypeptide chain, predominantly right-handed, with maximal intrachain hydrogen bonding of the peptide bonds; one of the most common secondary structures in proteins.

(A+T)/(G+C) ratio

a reference to the base composition of double-stranded DNA. DNA from different sources has different ratios of the A-to-T and G-to-C base pairs, e.g. DNAs isolated from organisms that live in hot springs have a higher GC content, which takes advantage of the increased thermal stability of the GC base pair.

3' end

the end of a nucleic acid that lacks a nucleotide bound at the 3' position of the terminal residue.

5' end

the end of a nucleic acid that lacks a nucleotide bound at the 5' hydroxyl group.

A form

a duplex DNA structure with right-handed twisting in which the planes of the base pairs are tilted about 70° with respect to the helix axis. A linear polymer of amino acids held together by peptide linkages. The polypeptide has a directional sense, with an amino- and a carboxy-terminal end.

a(n)-

without.

ab-

from.

abortive mitosis model

the hypothesis that cells undergo a programmed death when induced to mitosis at an inappropriate phase of the cell cycle.

absolute configuration

1. the actual, as opposed to relative to some other compound, orientation of atoms in space at an asymmetrical centre.

2. the configuration of four different substituent groups around an asymmetric carbon atom; for historical reasons molecules in biochemistry are

compared to D- and L-glyceraldehyde.

■ absorption

1. in immunology, the neutralisation of immune serum by some of the antigens against which it has been raised. Serum raised against a hapten attached to a larger protein may be exposed to the carrier protein alone in order to absorb the antibodies directed towards epitopes of the protein, leaving the serum with un-neutralised antibodies directed only towards the hapten.

2. transport of the products of digestion from the intestinal tract into the blood.

■ absorption spectrum

the molar absorption (extinction) coefficient as a function of wavelength, usually displayed with absorbance on the ordinate and wavelength on the abscissa.

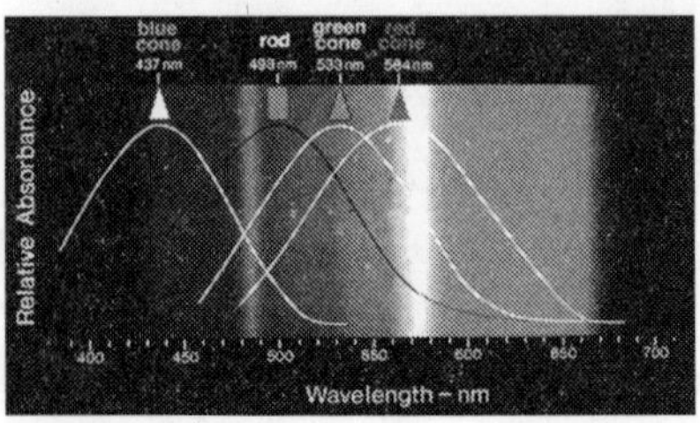

■ Abu shunt

an alternative pathway in bacteria and higher plants for entry of glutamate into the tricarboxylic acid cycle: glutamate γ-aminobutyrate (Abu) → succinic acid semialdehyde → succinate.

■ acceptor

in enzyme mechanisms, a functional group of an enzyme that transiently receives a moiety of a substrate, the *donor*, before itself becoming a donor in transferring it to a second substrate, which is also an 'acceptor'. More generally in immunology, pharmacology and cell biology, an acceptor is an entity that receives an atom, ligand or structure from a 'donor'.

■ acceptor (3') splice site

a site in pre-mRNA that corresponds to the 3'-end of the intron and the 5'-end of the next exon. The last two bases of the intron at the acceptor splice site are AG in most cases, but AC in some instances.

■ acceptor control (Electron transport chain)

the regulation of the rate of respiration by the availability of ADP as a phosphate group acceptor.

Electron Transport Chain

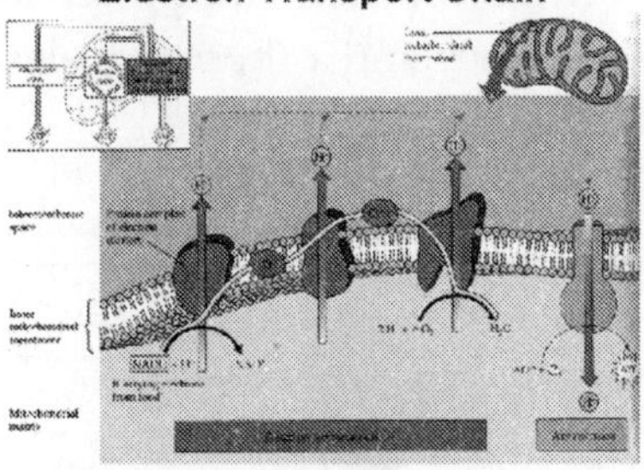

■ accessory pigments

visible light-absorbing pigments such as carotenoids and xanthophyll in green plants and photosynthetic bacteria that trap energy from sunlight and pass it on to "special pairs".

■ acetal

the product formed by the successive condensation of two alcohols with a single aldehyde. It contains two ether-linked oxygens attached to a central carbon atom.

■ acetate rule

the observation, first based on inspection of structures, later on experimentation, that many natural products appear to have been assembled from multiple acetate (acetyl-CoA and/or malonyl-CoA) units in head-to-tail condensations.

■ acetogenin

a compound derived from acetyl units donated by acetyl- and/or malonyl-CoA units, assembled into a non-reduced polyketide, i.e. with the carbonyl groups intact, often, then, cross-linked by aldol condensation and processed by further biochemical transformations to the final product, e.g. orsellinic acid, griseofulvin.

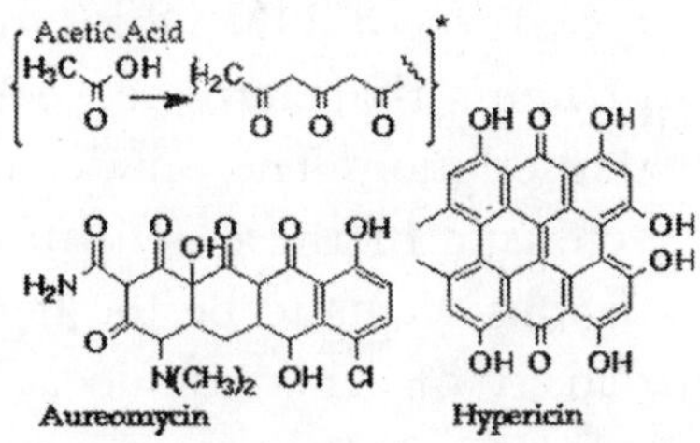

■ acetyl CoA

acetyl-coenzyme A, a high-energy ester of acetic acid that is important both in the tricarboxylic acid cycle and in fatty acid biosynthesis.

■ Achilles heel cleavage

a method for cleaving of DNA at very specific sequences. The site of cleavage is able to bind a specific protein, e.g. the *lac* repressor, that already contains a restriction site or is engineered to contain one; the presence of the DNA-binding protein protects the restriction site while the rest of the DNA is enzymically methylated, thus eliminating all unprotected restriction sites. Upon removal of the binding protein, the DNA can be cleaved at the one remaining restriction site.

■ acid blob

a feature of DNA-binding proteins that activate yeast transcription. The only characteristic of these activating regions seems to be the high density of acidic amino acid residues.

■ acidosis

a condition in which blood pH decreases, either for metabolic or respiratory reasons.

■ acinar cell

a secretory cell within an acinus.

■ acinus

a cluster of secretory cells surrounding a duct.

■ acrosome

a specialised lysosome of a spermatozoon that contains hyaluronidase, the proteinase acrosin and other hydrolytic enzymes.

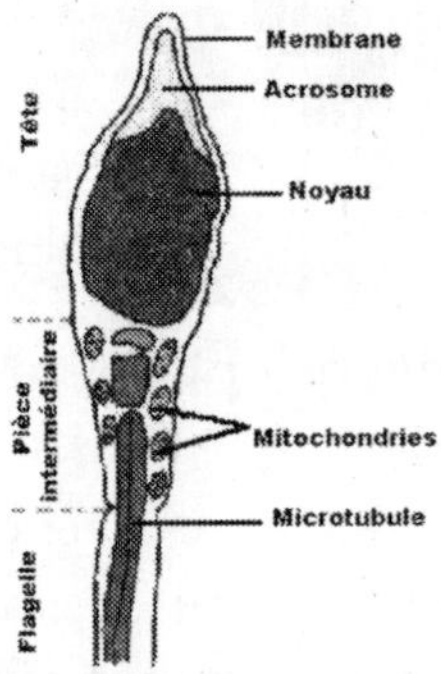

■ actin

a protein found in combination with myosin in muscle and also found as filaments constituting an important part of the cytoskeleton in many eukaryotic cells.

■ actinomycin D

an antibiotic that binds to DNA and inhibits RNA chain elongation.

■ action spectrum

a measure of the effectiveness of various wavelengths

in promoting photosynthesis; usually a plot of photosynthetic efficiency against wavelength.

■ activated complex

the highest free energy state of a complex in going from reactants to products.

■ activation

in immunology, the immune response of an organism to a foreign substance; lack of a response is called tolerance.

■ activation energy

1. the energy needed to raise the reactants, or an enzyme-substrate complex, to the transition state, where it has an equal likelihood of conversion to product or reversion to reactants. This value is commonly evaluated in an *Arrhenius plot*, ln*k* against 1/*T*, where *k* is the rate constant, *T* is the absolute temperature and the slope is E_a/R (E_a is the activation energy and *R* the gas constant).
2. ($\Delta G^{o\prime}$) The amount of energy (SI unit of joules) required to convert 1 mole of a reactant from the ground state to the transition state.

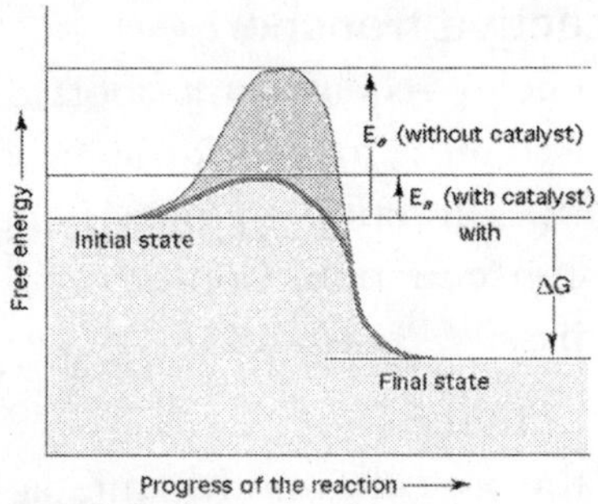

■ activator

in enzyme kinetics, a compound that increases the rate of an enzymic reaction.

■ active acetaldehyde

the CH_3CHOH- group when attached to the thiamin pyrophosphate prosthetic group of pyruvate decarboxylase in the pyruvate dehydrogenase complex.

■ active glycolaldehyde

the a-hydroxyethylidene group when attached to the thiamin pyrophosphate prosthetic group of transketolase.

■ active site

the binding and catalytic sites of an enzyme; more loosely, those residues of an enzyme that interact with a substrate or participate in any way in binding or catalysis.

■ active transport

energy-requiring transport of a solute across a membrane in the direction of increasing concentration. Contrast with passive transport.

■ activity

the true thermodynamic activity of a substance, as distinct from its molar concentration. Most of the time we call the activity equal to the molar concentration and ignore it.

■ acyl-enzyme

an intermediate in the hydrolysis of substrates by some peptidases and esterases, e.g. by serine proteinases, in which the acyl moiety of the substrate is transiently attached to a serine hydroxy group of the enzyme.

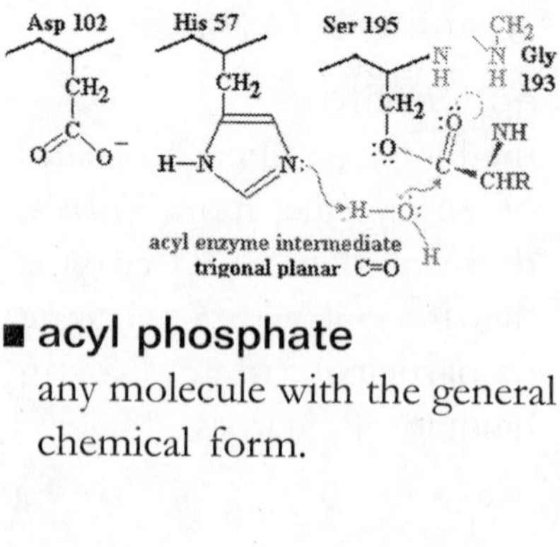

■ acyl phosphate

any molecule with the general chemical form.

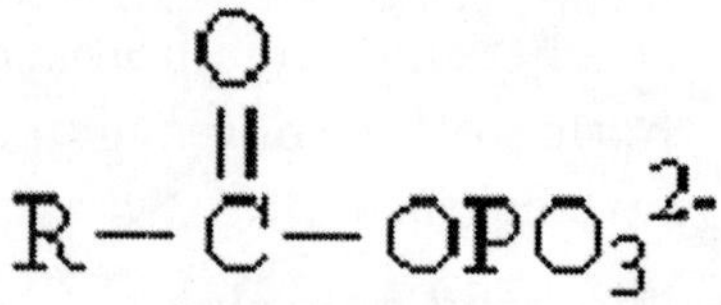

■ ad-

towards.

■ adaptation

the evolution of a feature or function through natural selection of incremental improvements, as contrasted with expectation.

■ adaptor hypothesis

the proposal, made before the roles of mRNA and tRNA in protein synthesis were established, that a low-molecular-mass form of RNA provides the interface between each amino acid and the template. The hypothesis assumed that the interaction of the adaptor and a polynucleotide template is much more specific than that between the amino acid and the template.

■ additivity

the principle of thermodynamics that free energy (or enthalpy of entropy) is a sum of contributions of independent components, e.g. the free energy of a protein transition is the sum of free energy changes owing to hydrogen bonding, electrostatic interactions, van der Waals packing effects, restriction of rotation about bonds, interactions with solvents, etc. The utility of this approach is undermined by the lack of a true independence of these factors, resulting in an inability to accurately evaluate their individual contributions to the overall change in free energy.

■ **adenine**
a purine base found in DNA or RNA.

■ **adenosine**
a purine nucleoside found in DNA, RNA, and many cofactors.

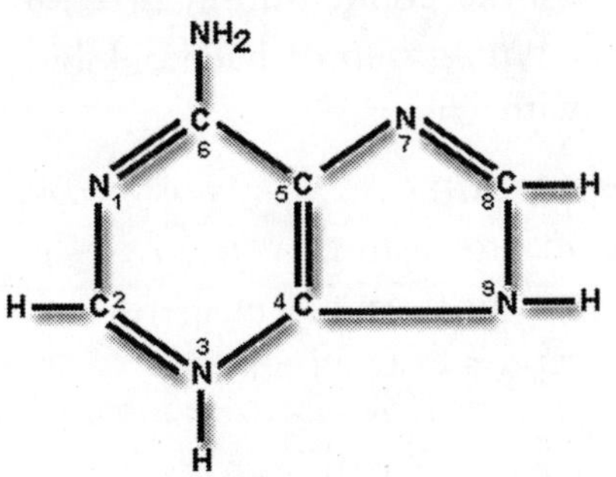

■ **adenosine 3',5'-cyclic monophosphate**
cAMP.

■ **adenosine diphosphate (ADP)**
a ribonucleoside 5'-diphosphate serving as phosphate group acceptor in the cell energy cycle.

■ **adenosine triphosphate (ATP)**
the nucleotide formed by adding yet another phosphate group to the pyrophosphate group on ADP.

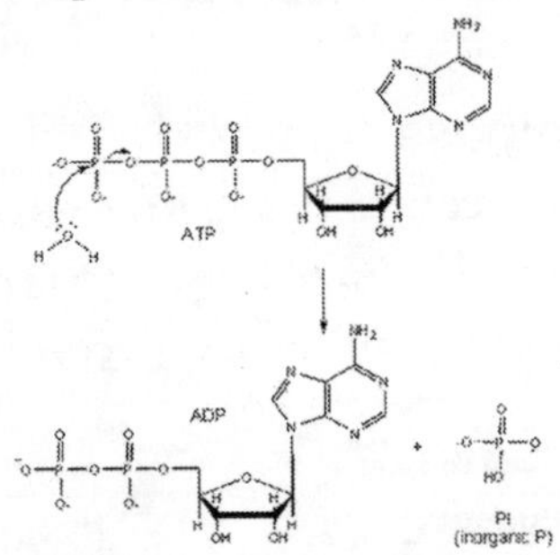

■ **adenylate cyclase**
the enzyme that catalyses the formation of cyclic 3',5' adenosine monophosphate (cAMP) from ATP.

■ **adenylate energy charge**
a measure, on a scale of 0-1, of the degree of phosphorylation of adenine nucleotides;

$([ATP]+{}^{1}/_{3}[ADP])/([ATP]+[ADP]+[AMP])$.

■ adipo
fat.

■ adipocyte
a specialised cell that functions as a storage depot for lipid.

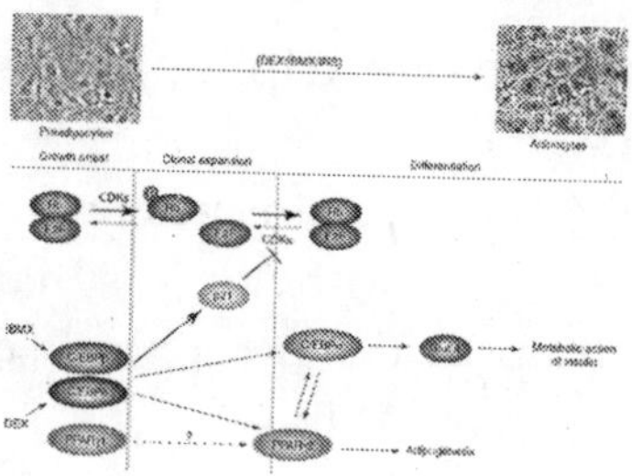

■ adipose tissue
connective tissue specialised for the storage of large amounts of triacylglycerols.

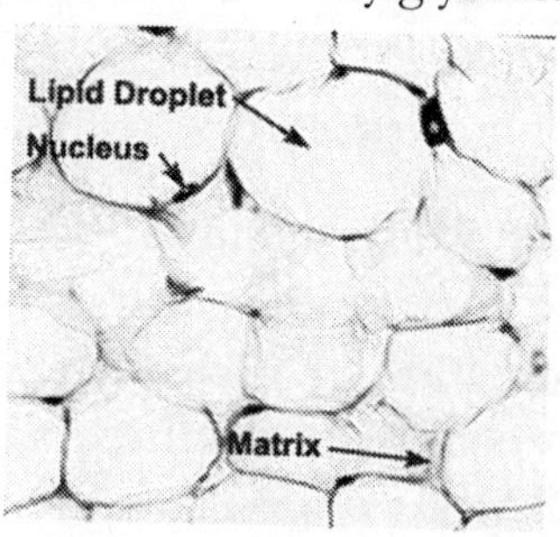

■ ADME-
an anachronism used in drug development, meaning absorption, distribution, metabolism and excretion.

■ A-DNA
a right-handed helix; a variant of the dominant B-form of DNA in solution, in which the base pairs are tilted somewhat out of the perpendicular orientation to the axis of the helix.

■ adrenergic
responsive to the adrenal medullary hormone, i.e. adrenaline (epinephrine), and by extension responsive to other catecholamines. $-Adrenergic responses are those that result in the intracellular generation of cyclic AMP.

■ advanced glycation end product (AGE)
the degradation product of a protein-carbohydrate addition product such as haemoglobin A_{1C}, the Schiff base formed by the condensation of an-g amino group of haemoglobin with glucose.

■ aerobe
an organism that uses oxygen as the terminal electron acceptor in respiration.

■ aerobic

requiring or occurring in the presence of oxygen.

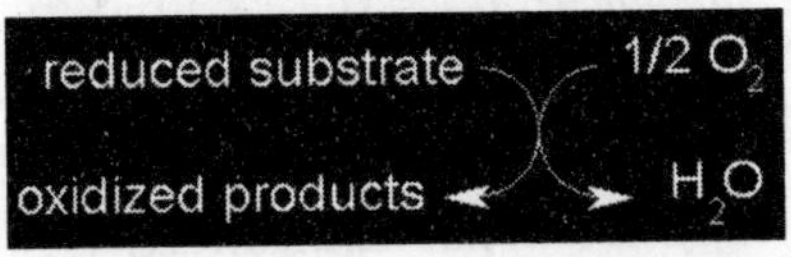

■ afferent

in neuroanatomy and physiology, this describes the neuron which delivers a signal to a synapse; contrasted with efferent, which refers to the neuron which receives and propagates a signal.

■ affinity chromatography

1. a column chromatographic technique that employs attached functional groups that have a specific affinity for sites on particular proteins.
2. the separation of soluble macromolecules by use of a stationary phase that is designed to interact specifically with, and thus retard the elution of, the desired material; e.g. a hapten attached to a resin to help isolate an immunoglobulin directed against it.

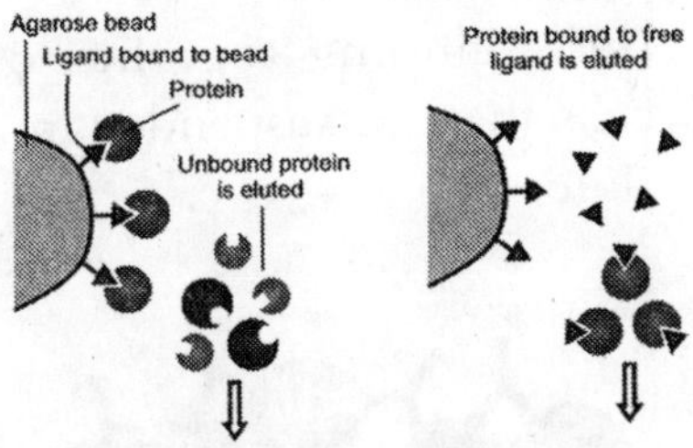

■ affinity labelling

a technique that depends upon the tight attachment of a ligand to a binding site on a protein or cell, followed by a chemical reaction to link the ligand covalently to its binding site.

■ affinity precipitation

a technique for purification of proteins that depends upon reversible attachment to a ligand. A *bis-ligand* (or *homofunctional ligand*) is a double-headed molecule that can attach at each end to part of a multisubunit protein, forming an insoluble lattice. In an alternative approach, a ligand attaches a protein to a water-soluble polymer that can be made insoluble by changing the conditions of pH, temperature or ionic strength.

■ agglutination

the clumping together of cells that are suspended in a fluid.

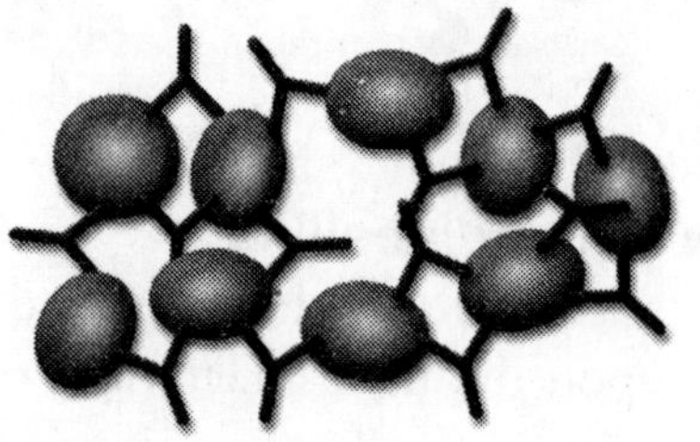

■ agglutinin

a compound that cross-links cells, e.g. a lectin, an antigen that reacts with sensitised cells.

■ aggregation

the formation of higher-molecular-mass species due to non-covalent adherence of smaller species. Especially for proteins, aggregation is a form of denaturation in which non-polar surfaces of secondary structures, e.g. those of a-helices and $-sheets that normally form intramolecular interactions and are buried within the interior of the protein, are allowed to interact intermolecularly and to form multimolecular forms that are sometimes insoluble. Aggregation is contrasted with *oligomerisation*, the normal interaction of native, correctly folded, proteins into higher-order multimers.

■ aglycone

the non-sugar moiety of a glycoside.

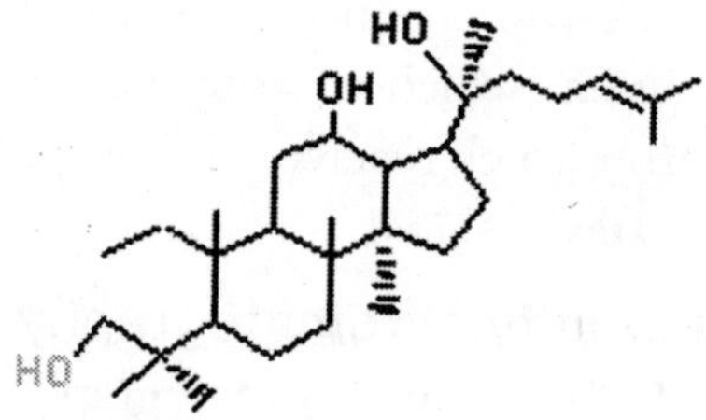

■ alarmone

a metabolite that signals a cell's response to changing environmental conditions, e.g. guanosine -3', 5'-bispyrophosphate in bacteria, diadenosine-5,5'-P1,P4-tetraphosphate in blood cells.

■ albumin

66.3 kDa protein comprising most of the protein in serum and largely responsible for buffering pH and volume of blood.

■ alcohol
a molecule with a hydroxyl group attached to a carbon atom.

■ alcohol fermentation
the anaerobic conversion of glucose to ethanol *via* anaerobic glycolysis.

■ aldehyde
a molecule containing a doubly bonded oxygen and a hydrogen attached to the same carbon atom.

■ aldose
a simple sugar in which the carbonyl carbon atom is an aldehyde.

■ alkaloids
nitrogen-containing organic compounds of plant origin. They are usually basic, and often have biological activity. An example would be caffeine.

■ alkalosis
a metabolic condition in which blood pH decreases, usually the result of a metabolic condition or vomiting.

■ alleles
alternative forms of a gene.

■ allo-
other.

■ allostasis
the altered state caused by stress, trauma, disease, aging, etc. The cumulative burden (stated in physiological, endocrinological or psychiatric terms, depending upon the context) of the shift from homeostasis to allostasis, is the allostatic load.

■ allosteric activator
any molecule which positively modulates the activity of an allosteric enzyme.

■ allosteric enzyme
a regulatory enzyme, with catalytic activity modulated by the noncovalent binding of a specific molecule at a site other than the active site.

■ allosteric site
the specific site on the surface of an allosteric enzyme molecule (distinct from the active site) to which a modulator molecule binds.

■ allotype
a classification of immunoglobulin molecules according to the antigenicity of the con-

stant regions; a variation that is determined by a single allele.

■ **alpha -helix**

a secondary structure in proteins; the right-handed helical folding of a polypeptide such that amide nitrogens share their hydrogen atoms with the carbonyl oxygens of the fourth amide bonds towards the C-terminal end of the polymer.

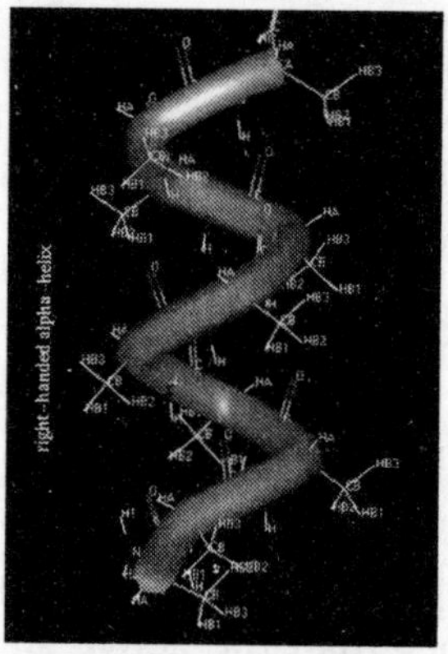

■ **alpha-imino acid**

a carboxylic acid that bears a substituted amino group on its α carbon; commonly proline, hydroxyproline, sarcosine.

■ **Alu sequence**

a small interespersed repeat element (SINE); a repetitive sequence in human DNA found in some introns, and characterised by Alu restriction sites; the locus of some homologous recombinations. Comprising 3-6% of the human genome, the sequences are about 300 bp long and usually feature a head-to-tail tandem repeat.

■ **ambi-**

both.

■ **ambidexteran**

an aggregate of glycosaminoglycans that, unless prevented by steric hindrance or over-or under-sulphation, presents one or both surfaces of their tape-like shape to the complementary surface of another; this results in linear polymers if aggregation is on only one surface, or sheets if aggregation can occur on both surfaces.

■ **Ames test (Salmonella test)**

a test for mutagenicity and carcinogenicity which uses specially constructed bacterial strains, by screening the effects of test compounds for their ability to produce reverse mutations that will restore the ability of the bacteria to grow in the absence of an essential metabolite. Named after it's inventor Bruce Ames.

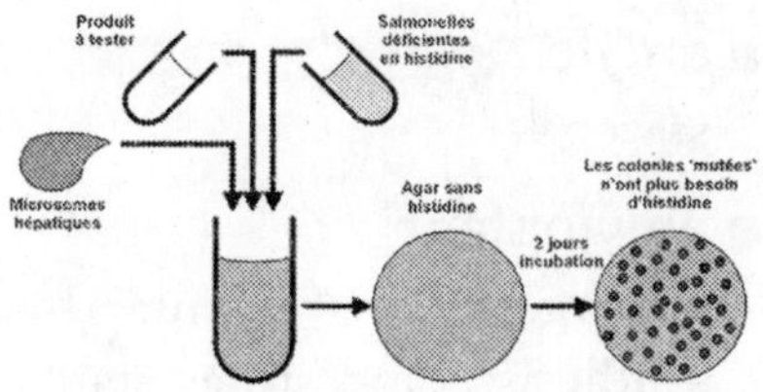

■ **aminimide**

a peptide analogue in which an amino nitrogen substitutes the α-amino group of an amino acid residue; this forms a stable, soluble product.

■ **amino acid activation**

enzymatic esterification of the carboxyl group of an amino acid to the 3'-hydroxyl group of its corresponding tRNA which requires 2 ATP-equivalents of energy.

■ **amino acids**

molecules which serve as the building blocks of proteins and having the general formula.

■ **aminoacyl-tRNA**

an aminoacyl ester of a tRNA.

■ **aminoacyl-tRNA synthetases**

enzymes that catalyse synthesis of an aminoacyl tRNA at the expense of ATP energy.

■ **amino-terminal residue**

the amino acid residue in a polypeptide chain which has a free α-amino group.

■ **aminotransferases**

enzymes that catalyse the transfer of amino groups from α-amino to β-keto acids; commonly called transaminases.

■ **ammonotelic**

organisms which excrete excess nitrogen in the form of ammonia.

■ **amphi-**

on both sides.

■ **amphibolic pathway**

a metabolic pathway used in both catabolism and anabolism.

■ **amphipathic**

having both polar and non-polar groups, e.g. a detergent.

■ **amphipathic helix**

a protein structure that serves in part as an interface between polar and non-polar phases; an a-helix that displays non-polar residues on one side and polar residues on the other (e.g. in many globular proteins).

■ **amphiprotic**
capable of donating and accepting protons, thus able to serve as an acid or a base.

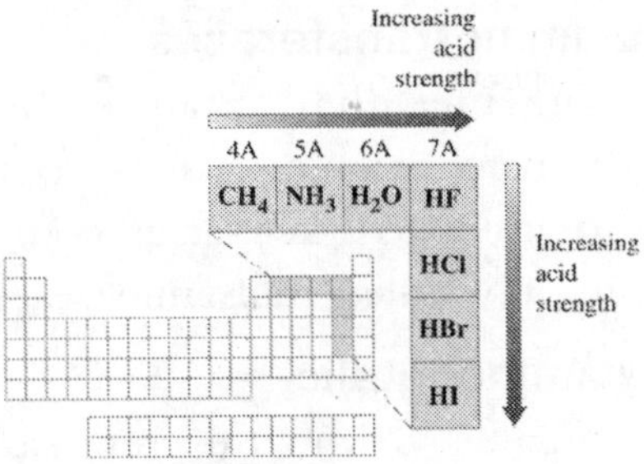

■ **amphitropic**
having an affinity for both lipid and aqueous environments, e.g. a membrane-associated protein that has domains that are embedded in the membrane and others that extend into the cytoplasm or the extracellular space.

■ **amphoteric**
having both acidic and basic groups, e.g. an amino acid.

■ **amplification**
the causation of a quantitatively major biochemical or physiological event by a quantitatively minor initiator, e.g. the blood clotting cascade, ion gating; also, using the polymerase chain reaction, the increase by orders of magnitude of copies of a DNA template.

■ **amylo**
starch.

■ **amyloplast**
an organelle of plants that synthesises and stores starch.

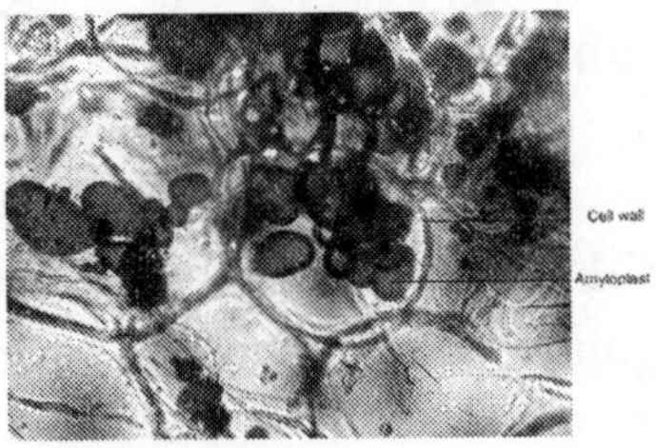

Potato Cells Showing Amyloplasts (400x)
Amyloplasts are a starch storing plastid.

■ **anabolism**
the phase of intermediary metabolism concerned with the energy-requiring biosynthesis of cell components from smaller precursors.

■ **anaerobe**
an organism that lives without oxygen and uses another chemical species as a terminal electron acceptor. Obligate anaerobes die when exposed to oxygen; facultative anaerobes can live in both

aerobic and anaerobic environments.

■ anaerobic

occurring in the absence of air or oxygen.

■ analytical ultracentrifugation

a technique of very-high-speed centrifugation that sediments soluble macromolecules and characterises them according to their rate of sedimentation (*sedimentation-velocity ultracentrifugation*) or the extent of their sedimentation (*equilibrium sedimentation ultracentrifugation*).

■ anaplerotic pathway

metabolic reactions that replenish the pools of intermediates of the tricarboxylic acid cycle. These pools may become depleted, as they also serve as precursors for amino acid synthesis, gluconeogenesis and other anabolic reactions.

■ anaplerotic reaction

an enzyme-catalysed reaction that can replenish the supply of intermediates in a metabolic pathway, most commonly the citric acid cycle.

■ anchored PCR

a variation of the PCR technique, similar to ligation-mediated PCR, that is applied to double-stranded DNA fragments for which the sequence at only one end of the gene is known. The technique allows amplification of a complete sequence of a gene when only the N-terminal sequence of a protein is known. A short polynucleotide of known sequence is ligated to the 3'-ends of the double-stranded DNA so that a primer complementary to it can be added, along with the primer determined from the partial protein sequence.

■ androgen

a compound, usually a steroid, that supports the development of male secondary sex characteristics, e.g. testosterone.

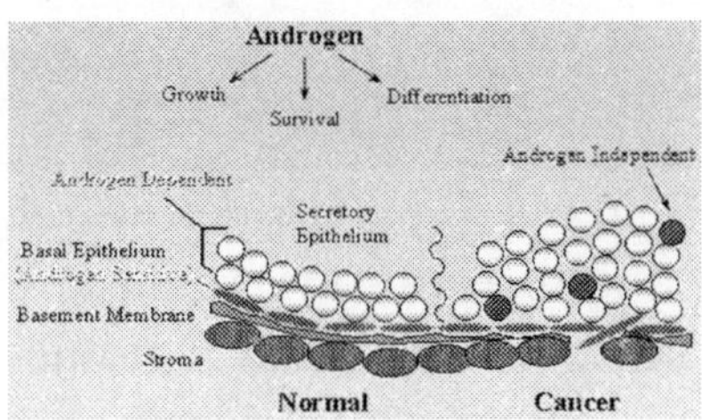

Anfinsen cage
a proposed model for chaperone-assisted protein folding in which a chaperone, GroEL for instance, provides an environment in which a denatured protein can unfold and correctly refold, undisturbed by the possibility of aggregation with other unfolded molecules.

angstrom (Å)
a unit of length (10^{-10} cm) sometimes used to indicate molecular dimensions.

anhydride
the product of the condensation of two carboxyl or phosphate groups in which the elements of water are eliminated to form a compound with the general structure shown below where X is either carbon or phosphorus (a phosphoanhydride).

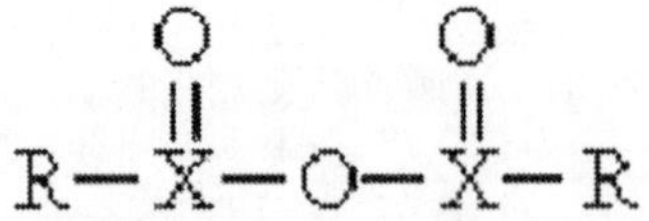

anion-exchange resin
a resin with fixed cationic groups which is used in the chromatographic separation of anions, e.g. DEAE (diethylaminoethyl)-cellulose.

anomers
the sugar isomers that differ in configuration about the carbonyl carbon atom. This carbon atom is called the anomeric carbon atom of the sugar.

ante-
before, forwards.

anthrone reaction
a colorimetric method for estimation of sugars that involves treatment with sulphuric acid and anthrone.

anti-
against, opposite.

antibiotic
a natural product that inhibits bacterial growth (is bacteriostatic) and sometimes results in bacterial death (is bacteriocidal).

antibody
a defense protein synthesised by the immune system of vertebrates. See also immunoglobulin.

■ **antibody**

a specific protein that interacts with a foreign substance (antigen) in a specific way.

■ **anticodon**

a sequence of three bases on the transfer RNA that pair with the bases in the corresponding codon on th messenger RNA.

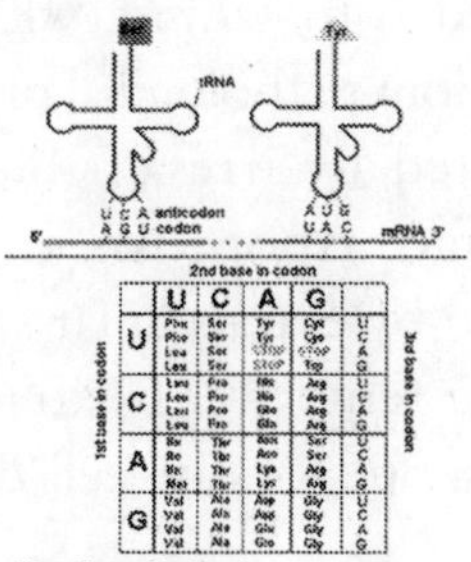

The Genetic Code

■ **antigen (immunogen)**

1. a substance that causes production of an antibody directed against itself.
2. any molecule which causes the synthesis of a specific antibody in vertebrates.

■ **antiparallel**

in protein chemistry, the orientation of extended polypeptide chains that interact in a pleated sheet structure, one chain in an N- to C-terminal direction and the other in a C- to N-terminal direction; in nucleic acid chemistry, the orientation of the two polynucleotide chains of a double helix, one that runs in a 3' to 5' direction and the other in a 5' to 3' direction.

■ **antiport**

cotransport of two solutes across a membrane in opposite directions.

■ **apical (luminal)**

descriptive of the free border of an epithelial cell, where it is in contact with vascular space.

■ **apicomplexan**

descriptive of a class of protozoan parasites that pass part of their life cycles in mammalian cells, e.g. the malaria pathogen *Plasmodium falciparum*, and which contain in their apices organelles that assist penetration of host cells. Also characteristic are the apicoplast, an organelle which contains maternally-inherited circular DNA, possibly of green algal evolutionary origin, and enzymes of the C6C3 metabolic pathway.

■ **apoactivator**

a regulatory protein that stimulates transcription from one or more genes in the presence of a coactivator molecule.

■ **apocrine**

descriptive of a secretion mechanism in which vesicles that contain the product burst through the cell membrane and are released along with some of the cytoplasm and plasma membrane.

■ **apoenzyme**

the protein portion of an enzyme, absent any organic or inorganic cofactors or prosthetic groups that might be required for catalytic activity.

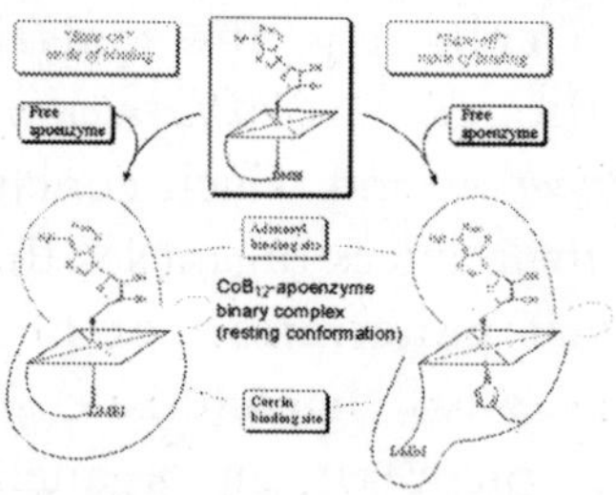

■ **apoprotein**

a protein stripped of any prosthetic group or metal ion normally associated with it.

■ **apoptosis**

a morphologically characterised process of programmed cell death, initiated by various physiological or pathological causes (e.g. cell turnover, hormone-induced atrophy, cell-mediated immune cytolysis, tumour regression), that is characterised by shrinkage of the nucleus and cytoplasm, cell fragmentation and phagocytosis. Apoptosis is contrasted with *necrosis*, which is a random pathological process initiated by irreversible cell damage. Apoptosis is controlled by extracellular signals or the removal of extracellular suppressors of cell death.

Intracellular Ca^{2+} and Apoptosis

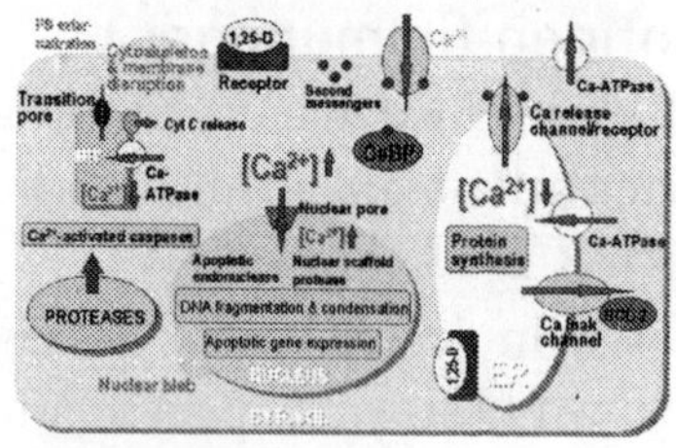

■ **aprotinin**

bovine pancreatic proteinase inhibitor.

■ **aptamer**

a polynucleotide with binding properties optimised for a particular ligand by cycles of selection from a mixture of random sequences.

- **apurinic DNA**
  a polynucleotide that has lost one or more purine bases due to the lability in acid of the glycosidic bond to purines.

- **apyridinic DNA**
  a polynucleotide that has lost one or more pyrimidine bases.

- **arthro**
  joint.

- **ascus**
  a spore-like form through which yeast cells pass during their life cycle; an outer wall surrounds a diploid cell that undergoes meiosis to form four haploid cells (ascospores) which eventually rupture to yield two diploid daughters.

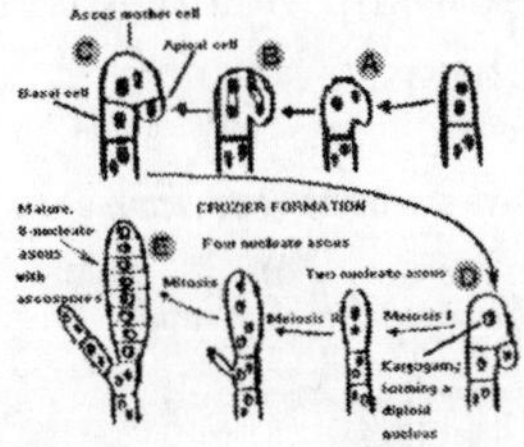

- **-ase**
  enzyme.

- **asexual reproduction**
  growth and cell duplication that does not involve the union of nuclei from cells of opposite mating types.

- **asymmetric carbon atom**
  a carbon atom that is covalently bonded to four different groups and thus may exist in two different stereospecific configurations (D- and L- in biochemistry).

- **-ate**
  negatively charged form of an acid.

- **ATP synthase**
  an enzyme complex that forms ATP from ADP and phosphate during oxidative phosphorylation oin the inner mitochondrial membrane or the bacterial plasma membrane, and during photophosphorylation in chloroplasts. Uses a proton gradient to

chemiosmotically drive the synthesis.

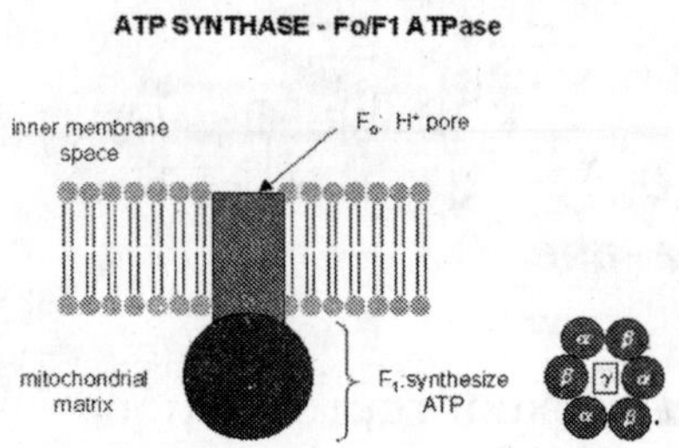

■ ATPase

an enzyme that hydrolyses ATP to yield ADP and phosphate; usually coupled to some process requiring energy such as the sodium potassium ATPase.

■ atrophy

the wasting away of an organ and/or its capabilities.

■ attenuator

an RNA sequence involved in regulating the expression of some prokaryotic genes.

■ a-type ion

in mass spectrometry, an N-terminal fragment of a polypeptide produced by cleavage of the bond between an α-carbon and its neighbouring α-carboxy carbon. A *b-type ion* is an α-N-terminal acylium ion which results from cleavage of an amide bond, and *y-type ions* are C-terminal iminium ions from such a cleavage.

■ autocrine

descriptive of a secretion that binds to receptors on the surface of the same cell that produces it.

■ autophagocytosis

the response of a cell to stress in which it forms an *autophagosome*, an organelle that contains parts of the cell's own cytoplasm. Later this will fuse with a lysosome to form an *autolysosome*; all such inclusions that derive from autophagocytosis are termed *autophagic vesicles.*

■ autoradiography

the technique of exposing film in the presence of disintegrating radioactive particles. Used to obtain information on the distribution of radioactivity in a gel or a thin cell section.

■ autoregulation

the process in which a gene regulates its own expression.

■ **autosome**

a chromosome that is neither of the sex chromosomes, e.g. in humans, a chromosome that is neither an X or Y.

■ **autotroph**

an organism that can synthesise its own complex molecules from simple carbon and nitrogen sources, such as CO_2 and NH_3.

■ **auxin**

a plant growth hormone usually concentrated in the apical bud.

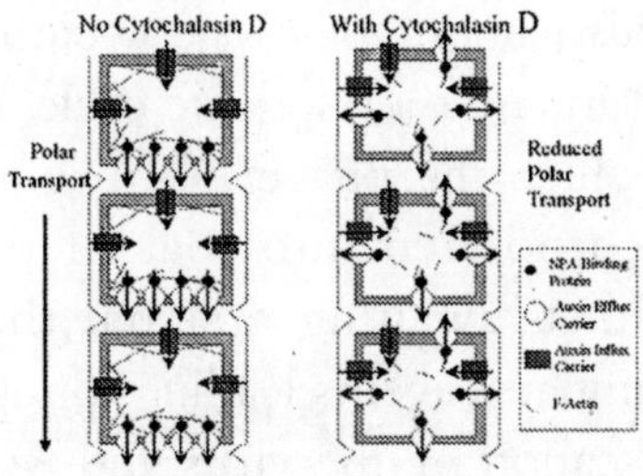

■ **auxotroph**

a mutant that cannot grow on the minimal medium on which a wild-type member of the same species can grow.

■ **auxotrophic mutant**

(auxotroph) A mutant organism defective in the synthesis of a given biomolecule, which must therefore be supplied for the organism's growth.

■ **Avogadro's number**

the number of molecules in a gram molecular weight of any compound (6.023×10^{23}).

■ **axon**

a long projection of a neuron through which it communicates with other cells.

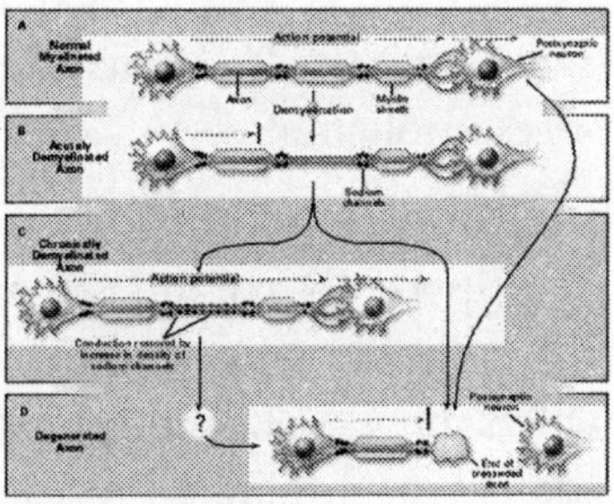

■ **axoplasm**

the cytoplasm of an axon.

■ **B cell**

one of the major types of cells in the immune system.B cells can differentiate to form memory cells or antibody-forming cells.

■ **B form**

the most common form of duplex DNA, containing a right-handed helix and about 10 (10.5 exactly) base pairs per turn of the helix axis.

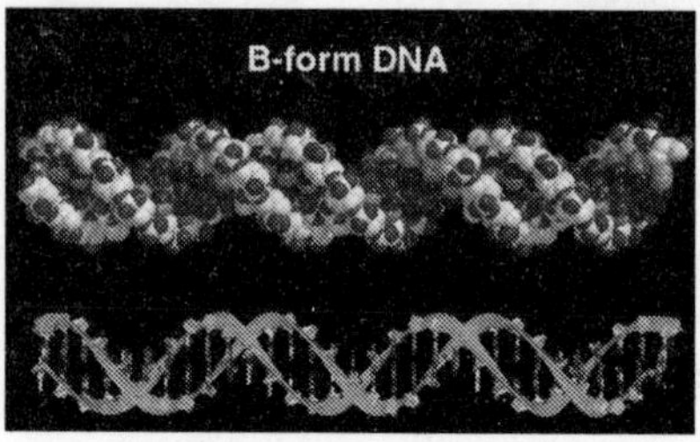

■ **backbone**

in chemistry, the longest continuous chain of atoms bonded to each other, exclusive of all others, that comprise a polymer.

■ **backbone cyclisation**

an approach to design of polypeptide-like drugs, in which the C- or N-termini or side chains are linked to each other (end-to-end, side chain-to-side chain, or end-to-side chain) in order to impose conformational restraints and thus limit proteolytic degradation and enhance binding to a target site.

■ **back-mutation**

a mutation that causes a mutant gene to revert back to its wild-type genotype.

■ **bacterial artificial chromosome (BAC)**

a cloning vector that can accept up to 350-kb fragments for cloning and sequencing of fragments of the human genome. Phage artificial chromosomes are similarly used.

■ **bacteriophage**

a virus that infects bacteria. Many phage have proved useful in the study of molecular biology and as vectors for the transfer of genetic information between cells. *Lytic phage*, e.g. the T series phage that infect *E. coli* (*coliphages*), invariably lyse a cell following infection; *temperate phage*, e.g. lambda bacteriophage, can also undergo a lytic cycle or can enter a lysogenic cycle, in which the phage DNA is incorporated into that of the host, awaiting a signal that initiates events leading to replication of the virus and lysis of the host cell.

■ **balanced polymorphism**

the result of selective pressures for and against a deleterious mutation that permits it to persist in a population. An example is the stable presence in Africa of the sickle cell gene due to the protection against malaria enjoyed by heterozygotes, i.e. those with sickle cell trait.

■ **basal metabolic rate**
the rate of oxygen consumption by an animal's body at complete rest under fasting conditions.

■ **base**
in nucleic acid chemistry, one of the nitrogenous compounds, i.e. purines and pyrimidines, that are incorporated into nucleosides, nucleotides and nucleic acids. The most common bases are adenine, cytosine, guanine, thymine and uracil, abbreviated as A, C, G, T and U respectively.

■ **base analog**
a compound, usually a purine or a pyrimidine, that differs somewhat from a normal nucleic acid base.

■ **base flipping**
the distortion of a double-stranded DNA structure that disrupts a base pair and redirects one nucleoside of the pair outwards, where it can interact with a DNA-modifying enzyme such as a methyltransferase.

■ **base pair**
in a nucleic acid double helix, a purine and a pyrimidine on different strands that interact by hydrogen bonding, most commonly a GC or AT pair.

■ **base roll**
variation in orientation of bases in a DNA double helix that permits some tilting of the bases.

■ **base stacking**
the close packing of the planes of base pairs, commonly found in DNA and RNA structures.

■ **basement membrane**
an extracellular network of fibres and glycoconjugates that underlies and strengthens some tissues; an interface between these tissues and the connective tissue that surrounds them.

■ **basolateral**
descriptive of the border of an epithelial cell that is attached to the basement membrane.

■ **basophil**
a polymorphonuclear leucocyte containing granules that

react with a histological stain for basic substances.

Basophils

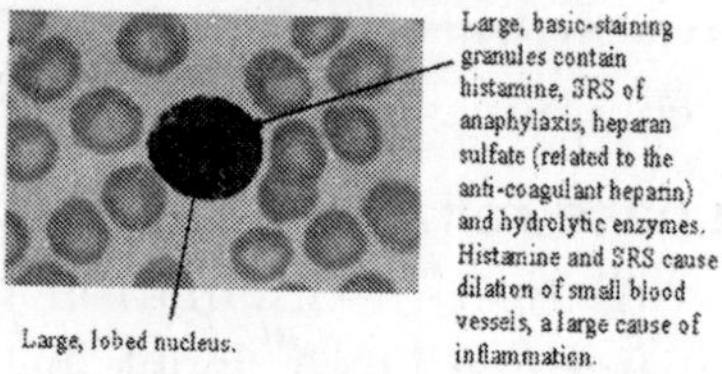

■ bathochromic shift

a shift to longer wavelengths.

■ B-cell (B-lymphocyte)

a lymphocyte that can be activated to proliferate and to synthesise and secrete a specific immunoglobulin G by binding of the antigen that is recognised by the immunoglobulin G receptor on the cell.

■ B-DNA

a right-handed helix; the dominant conformational variant of DNA in solution, in which the base pairs are stacked nearly perpendicular to the axis of the helix.

■ Beer's law

a quantitative treatment of the absorption of monochromatic light by a solution. The equation $A=\log(I_0/I)$ defines A (the absorbance, which is proportional to the concentration of the chromophore), I_0 (the intensity of the incident light) and I (the intensity of light having passed through 1cm of the solution). T, the transmission, is defined as $A=-\log T$.

■ Bence Jones protein

the immunoglobulin light chains that are synthesised in large amounts and are secreted into the urine by multiple myeloma patients.

■ beta -amino acid

a zwitterionic compound with carboxyl and amino groups on adjacent carbons: H_2N-CHR-CHR'-COOH.

■ beta barrel

a form of supersecondary structure: a structure within a globular protein in which six or more extended polypeptide strands (-pleated sheets) are arrayed cylindrically to form the staves of a barrel.

■ beta-bend

a characteristic way of turning an extended polypeptide chain in a different direction, involving the minimum number of residues, and held together by hydrogen bonding.

■ **beta-oxidation**
oxidative degradation of fatty acids that occurs by the successive oxidation of the β-carbon atom.

■ **beta-sheet**
a sheetlike structure formed by the interaction between two or more extended polypeptide chains.

■ **bi(s)-**
twice, double.

■ **bicistronic**
having two cistrons in tandem, e.g. the A and B subunits for the *Escherichia coli* enterotoxin.

■ **bidirectional replication**
synthesis of DNA that is effected by two replication forks that travel away from a single origin of replication.

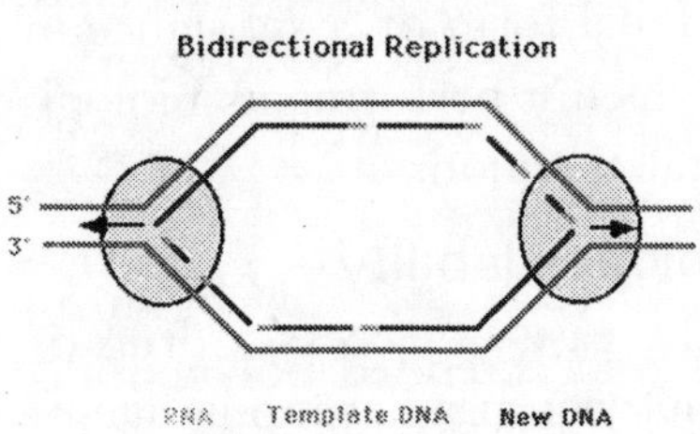

■ **bilayer**
a double layer of amphipathic lipid molecules which orient themselves so the hydrocarbon tails face inward to form a continuous nonpolar phase and the polar head groups face outward. In this way they form the basic structure of membranes.

■ **bile acid**
one of the products of cholesterol hydroxylation and side-chain oxidation to the level of a carboxylic acid. The carboxylate is often conjugated through an amide bond to a glycine or a cysteic acid. Excreted into the small intestine from the gall bladder, bile acids act as detergents, and aid lipid absorption.

■ **bile pigment**
one of the highly coloured products of haem degradation.

■ **bile salts**
amphipathic steroid derivatives with detergent properties, participating in digestion and absorption of lipids.

■ bilin

a tetrapyrrole pigment chemically related to the bile pigments, e.g. phycobilin, a bilin of red or blue-green algae.

■ binary system

in transgenic research, an approach to control expression of one transgene by a second, each initially established in its own pedigree. By crossing the two lines, doubly transgenic animals are created in which the control may become operational. The gene of interest is regulated by an exogenous ligand acting either as a positive regulator that binds to a repressor or as a negative regulator that binds to a transactivator. The repressor or transactivator are products of the second transgene.

■ binding energy

the energy derived from noncovalent interactions between enzyme and substrate or receptor and ligand, resulting from displacement of water, formation of hydrogen bonds, etc.

■ binding protein

a circulating protein that carries its ligand from one site in the body to another, e.g. thyroxine-binding protein; also any protein specialised for binding a ligand, e.g. a calcium-binding protein.

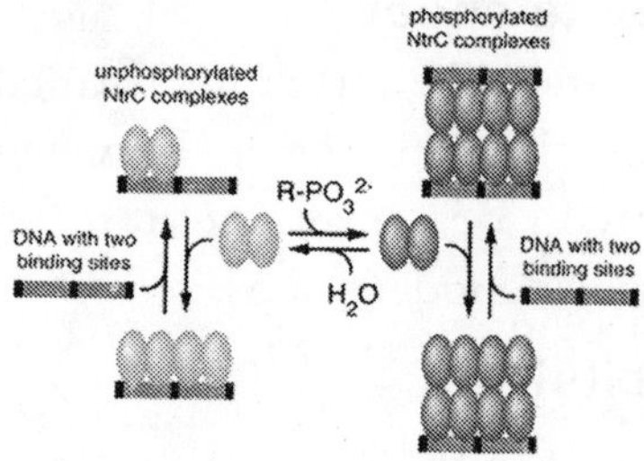

■ binding site

that region of the surface of an enzyme (or receptor, or binding or transport protein) that holds the substrate or product (or other ligand).

■ bio-

life.

■ bioactive

descriptive of the chemical, stereochemical or conformational form of a substance in which it performs its biological function.

■ bioavailability

the ease of access a drug or nutrient may have to its site of biological action; e.g. the ability of an antisense oligonucleotide designed to target a neuroreceptor to cross the blood-brain barrier, the ability of vitamin B12 to be absorbed by the digestive system.

■ biochemical genomics

an experimental approach to identification of the gene responsible for a given enzymic activity. Each open reading frame (ORF) of an organism, yeast for instance, is fused to the cDNA of glutathione S-transferase (GST) in a plasmid and then introduced into yeast. Yeast strains containing these plasmids are assayed in groups for the desired activity until the one strain with the activity is identified. (GST facilitates purification of the GST-ORF fusion protein by affinity chromatography.) Once the strain containing the activity is known, the ORF in the plasmid, which is responsible for the activity, is easily determined.

■ biochemical pathway

a series of enzyme-catalysed reactions that results in the conversion of a precursor molecule into a product molecule.

■ biocytin

the molecule arising from covalent attachment of biotin to a Lys residue *via* an amide linkage.

Biotin Binding to Map Topology and Surface Accesibility

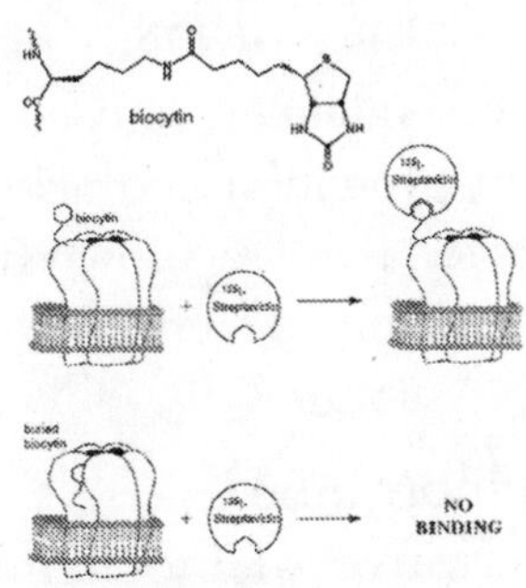

Probing the Major Immunogenic Region (MIR) - Site of the Autoimmune Disease Myasthenia Gravis

■ bio-imprinting

the induction of a new catalytic specificity in an enzyme in an organic solvent. An enzyme and a weakly binding non-substrate/non-product (e.g. chymotrypsin and *N*-acetyl-D-tryptophan) are suddenly precipitated together from an aqueous solution by addition of a miscible organic solvent. The complex reverts to the native conformation and behaviour when redissolved in water, but if redissolved in a suitable organic solvent it will be found to possess a new enzymic activity, e.g. for synthesis of *N*-acetyl-D-tryptophan ethyl ester from *N*-acetyl-D-tryptophan and ethanol.

■ **bioinformatics**
the management, manipulation and use of DNA nucleotide sequences, and consequent protein amino acid sequences, that are known from sequencing of genomes and cDNA libraries.

■ **bio-inorganic**
descriptive of biochemical materials that include elements, frequently metals, not popularly associated with them, e.g. Fe-S redox proteins, cobalamin, Zn fingers, the Fe-Mo centre of nitrogenase.

■ **bioluminescence**
the chemiluminescent emission of light by a living thing, e.g. firefly, certain fungi.

■ **biomimetic**
descriptive of a compound or process that artificially simulates the function of a natural compound or process, e.g. a catalyst that employs the principles of enzyme specificity and rate acceleration.

■ **biomimicry**
the design of structural materials according to examples from biology, e.g. a polymer strengthened by carbon fibres, inspired by the insect exoskeleton, which is a protein matrix reinforced by chitin fibres.

■ **biomolecular interaction analysis (BIA)**
the detection by a biosensor of a specific binding event for quantification of the ligand. One kind of BIA is *surface plasmon resonance*, the observation of polarisation of light reflected from a surface coated with one of the binding partners. The change in polarisation provides, in real time, a measure of binding of the second partner.

■ **biopterin**
an enzymatic cofactor derived from pterin and in-

volved in certain oxidation-reduction reactions.

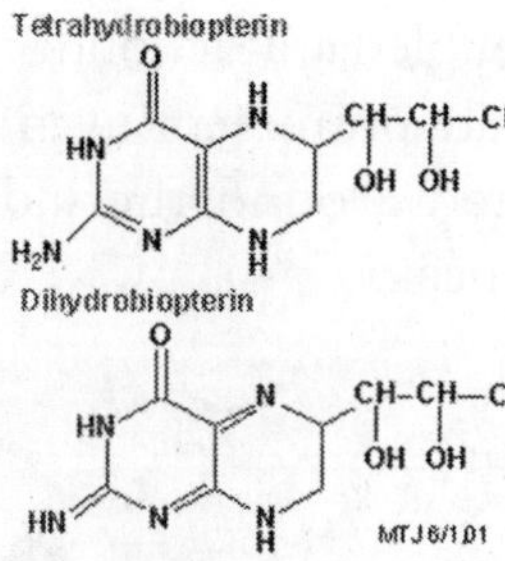

■ biosensor

a device, especially an electrochemical device, that detects some biological event (e.g. respiration, enzymic activity, binding to an antibody) and converts it into an electrical signal that it reports quantitatively and in real time.

■ biosphere

the zone of the earth in which life of some form is possible. It consists of the earth's surface, the oceans and atmosphere, and the upper several kilometers of the earth's crust.

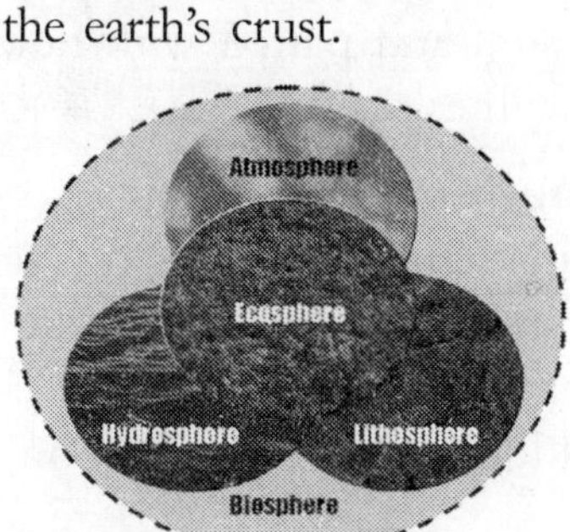

■ biosynthesis

the process by which a biological structure, especially a relatively simple structure, is formed by a sequence of enzymic reactions that starts from common metabolites or '*synthons*', e.g. the synthesis of haem from glycine and succinyl-CoA; the synthesis of steroids from acetyl-CoA. For the assembly of larger structures, such as membranes, via non-covalent interactions, the term *synkinesis* has been coined, with the subunits being termed *synkinons*.

■ biotin

an enzymatic cofactor involved in carboxylation reactions.

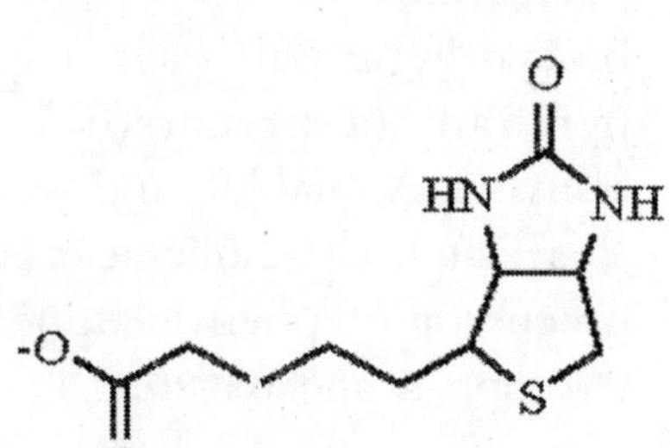

Biotin

■ blastoderm

the stage in embryogenesis when a unicellular layer at the surface surrounds the yolk mass.

■ **bleb**

a small bud-like projection from the surface of a cell.

■ **blunt ended (flush ended)**

descriptive of the structure of double-stranded DNA in which neither strand of the duplex extends further from the end than the other; often the product of cleavage by a restriction endonuclease.

■ **bond energy**

the energy required to break a bond.

■ **bovine**

derived from cattle.

■ **branch-point**

a single metabolite that is an intermediate in two or more biosynthetic pathways, e.g. pyruvate (a precursor of acetyl-CoA, alanine and oxaloacetate), chorismic acid (a precursor of phenylalanine, tryptophan and tyrosine).

■ **bricolage**

in evolutionary theory, the hit-or-miss recruitment and adaptation of pre-existing genes or materials for new functions.

■ **brown adipose tissue**

thermogenic fatty tissue with a high content of relatively uncoupled mitochondria; especially prominent in infants, located around the kidneys and neck.

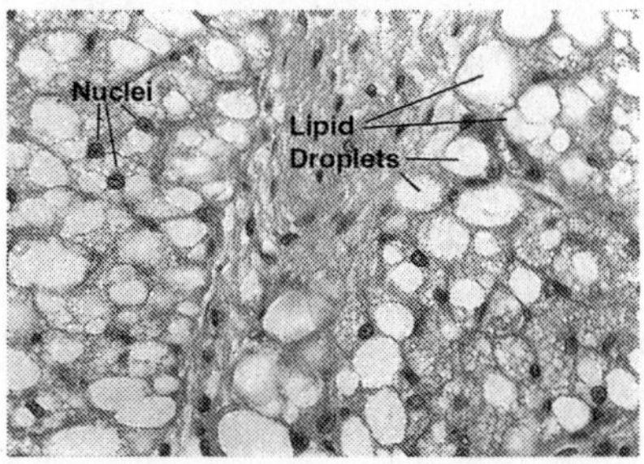

■ **bubble**

a widened segment of duplicating double-stranded DNA, visualised by electron microscopy, that is formed by two replication forks that travel away from the common origin of replication. The enlargement, or bubble, is a region in which the strands of the original DNA are separated and paired with newly synthesised DNA.

■ **buffer**

a system capable of resisting changes in pH when small amounts of acid or base are added. Consists of a conjugate acid-base pair in which the ratio of proton acceptor to proton donor is near unity.

■ **buoyancy factor**

in analytical ultra-centrifugation, one of the factors that determine a macromolecule's rate of sedimentation; (1-ur), where u is the partial specific volume of the macromolecule (expressed as ml/g) and r is the density of the solution (expressed as g/ml).

■ **C value**

the calculated value of the total DNA content of a cell per haploid number of chromosomes.

■ **c(o)ele**

cavity, hollow.

■ **CAAT box**

a regulatory sequence upstream from some eukaryotic structural genes.

■ **Caenorhabitis elegans**

a nematode; a small worm favored by developmental and molecular biologists because if its ability to grow under laboratory conditions, its short generation time and its transparency. Because of its simplicity, it is possible to trace the development of the zygote to each of the approx. 1000 cells of the adult.

■ **calorie**

the amount of heat required to raise the temperature of 1.0 g of water from 14.5 to 15.5 °C. One calorie (cal) equals 4.18 joules (J). Contrast with Cal, used in dietary applications, which is 1,000 cal.

■ **calpain**

a calcium-dependent cellular proteinase with a neutral pH optimum.

■ **calpastatin**

a protein inhibitor of a calpain.

■ **Calvin cycle**

the cyclic pathway used by plants to fix carbon dioxide and produce triose phosphates. Named after Melvin Calvin, an early worker in the field.

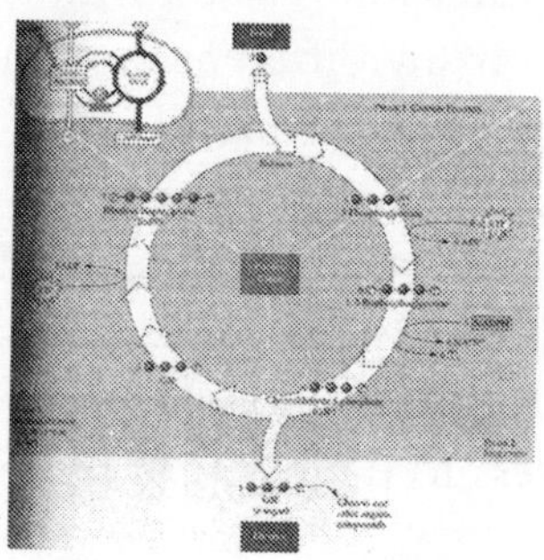

cAMP

a second messenger within cells; its formation by adenylate cyclase is stimulated by certain hormones (epinephrine) or other molecular signals. structure.

candidate

a compound, metabolite or peptide, etc. with a putative role; e.g. a candidate hormone may be a natural peptide which has in vitro bioactivity, but for which its characteristics in vivo have not yet been demonstrated.

canine

derived from the dog.

canyon hypothesis

a proposal to reconcile both the necessary invariability, and apparent non-immunogenicity, of the host cell receptor attachment site of a virus (e.g. a rhinovirus), and the ability of the same virus to mutate into many different immunologically recognised serotypes. The critical attachment site of the virus is sequestered at the bottom of a deep, narrow cleft, a 'canyon', which allows access to the receptor but not to an immunoglobulin; sites at the 'rim' of the canyon are accessible to immunoglobulins, but as they are not critical to the virus, they can mutate freely.

CAP

1. the 7-methylguanine nucleoside attached to the 5'-end of mRNA by a 5'-5'-triphosphate bond.

2. the catabolite gene activator protein, sometimes incorrectly referred to as the CRP protein. The latter term, in small letters (crp), should be used to refer to the gene but not to the protein.

cap site

the site on a DNA template where transcription begins. It corresponds to the nucleotide at the 5'-end of the RNA transcript which accepts the 7-methylguanine cap.

capping

covalent modification involving the addition of a modified guanidine group in a 5'-5" linkage. It occurs only in eukaryotes, primarily on mRNA molecules.

■ **caprine**
derived from the goat.

■ **capsid**
the protein coat of a virus particle.

■ **carbanion**
a transient species in which a carbon atom bears a formal negative charge, e.g. an intermediate in the carboxylation of glutamate residues of preprothrombin.

■ **carbocation (carbonium ion)**
any stable or transitory ion in which a carbon atom bears a formal positive charge.

■ **carbohydrate**
one of a class of biological materials comprising sugars, polymers of sugars, and compounds related to them. The name derives from the basic sugar structure, $(CH_2O)_n$. The category includes reduction and oxidation products, phosphate and sulphate esters, and amine derivatives.

■ **carbon cycle**
the movement of carbon atoms through different chemical forms and locations, from dissolved CO_2 in equilibrium with atmospheric CO_2 through plant carbohydrate, fats and proteins of plants and animals, and via oxidation back to atmospheric CO_2.

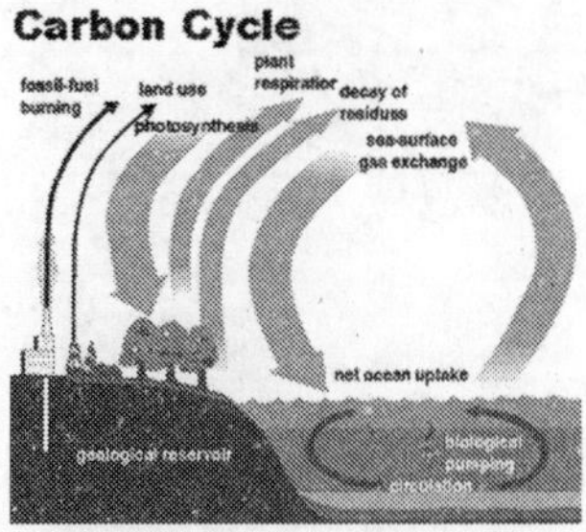

■ **carbon fixation reactions (dark reactions)**
the light-independent enzymatic reactions involved in the synthesis of glucose from CO_2, ATP, and NADPH.

■ **carboxylic acid**
a molecule containing a carbon atom attached to a hydroxyl group and to an oxygen atom by a double bond.

■ **carboxyl-terminal residue**
the amino acid residue in a polypeptide chain with a free α-carboxyl group.

■ **carcinogen**
a chemical that can cause cancer.

■ **carcinoma**
a malignant tumour of mesodermal origin.

■ **cardiac muscle**
a form of striated muscle characteristic of the heart.

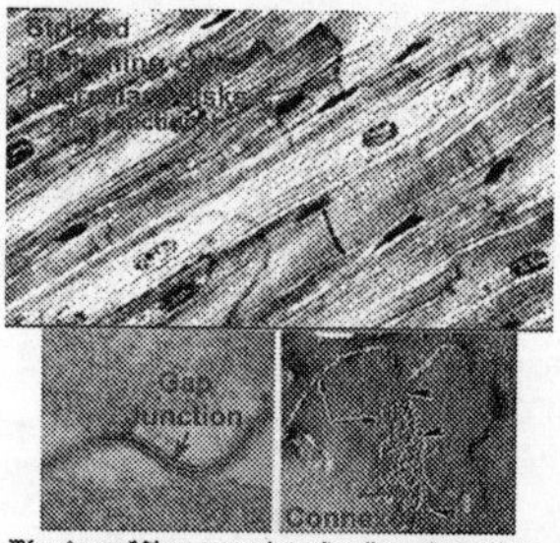

Electron Micrographs of a Gap Junction

■ **cargo**
a sub-cellular structure that is transported by an intracellular trafficking system, especially secretory or membrane vesicles along microtubules or chromosomes along the mitotic spindle.

■ **carotenoids**
lipid-soluble photosynthetic pigments made up of isoprene units. A derivative of phytoene, a symmetrical noncyclic 40-carbon terpene.

■ **carrier**
non-isotopically labelled material that dilutes a labelled tracer and enlarges the pool of the labelled metabolite, thus assisting in its isolation.

■ **cascade**
a series of enzymic reactions that at each step convert an inactive enzyme into an active enzyme, which in turn activates another inactive enzyme, and thus greatly amplifies the initial signal. In a homocascade all the events involve one enzymic type (e.g. proteolysis); in a heterocascade the events involve different enzymic types (e.g. phospholipase and kinase).

■ **caspase**
cysteinyl aspartate-specific proteinase; a member of a group of cellular cysteine endopeptidases that cleave so as to leave an N-terminal aspartate residue. The group includes interleukin-converting enzymes and enzymes associated with activation of apoptosis either as initiators near the head of a cascade of events, or as executioners, which perform limited proteolyisis of cellular proteins at the bottom of the cascade.

■ cata-
down.

■ catabolism
the action of energy-yielding metabolic pathways that degrade macromolecules and complex compounds or small molecules into CO_2, H_2O, etc.

■ catabolite gene activator protein (CAP)
a specific regulatory protein that controls initiation of transcription of the genes producing the enzymes required for a bacterial cell to use some other nutrient when glucose is lacking.

■ catabolite repression
the general repression of transcription of genes associated with catabolism that is seen in the presence of glucose.

■ catalyst
a compound that lowers the activation energy of a reaction without itself being consumed.

■ catalytic antibody (abzyme)
an antibody with catalytic properties; often raised against a hapten that chemically resembles the transition state of the intended substrate so as to force a substrate into that transition state.

■ catalytic site
the region of an enzyme that interacts with the substrate to effect the enzymic reaction.

■ catecholamine
one of a family of phenolic compounds chemically related to catechol (1,2-dihydroxybenzene), which is derived metabolically from tyrosine; the family comprises hormones and neurotransmitters, including adrenaline (epinephrine), noradrenaline, dopamine, etc.

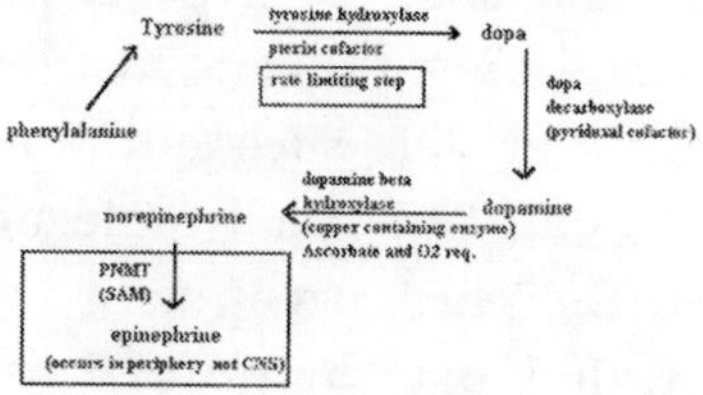

■ catenane
an interlocked pair of circular structures, such as covalently closed DNA molecules.

catenation

the linking of molecules without any direct covalent bonding between them, as when two circular DNA molecules interlock like the links in a chain.

cathepsin

an endopeptidase, often with low pH optimum, associated with lysosomes in the cell.

cation-exchange resin

a polymeric substance with fixed negative charges used in the chromatographic separation of cationic substances, e.g. CM (carboxymethyl)-cellulose.

caveolae

small invaginations of the plasma membrane, characterised by glypiated proteins among others, that provide binding sites and points of entry of small molecules into the cytoplasm via trancytosis or potocytosis. Trancytosis is the pinching off of the ligand-bearing vesicle into the cytoplasm. Potocytosis is the putative closure of the narrow neck of the invagination and consequent isolation of the vesicular space from the extracellular fluid; the high local concentration of the ligand allows diffusion of trapped molecules into the cytoplasm.

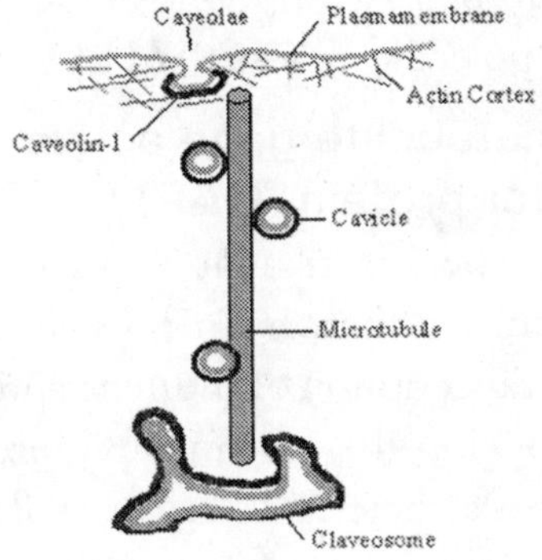

cDNA

complementary DNA, made in vitro from the mRNA by the enzyme reverse transcriptase using deoxyribonucleotide triphosphates. Unlike mRNA, cDNA can be easily propagated and sequenced.

cecropin

one of a group of inducible anti-microbial oligopeptides found in the haemolymph of a moth that serve some of the same functions as the immune system of higher species.

■ **cell commitment**

that stage in a cell's life when it be comes committed to a certain line of development.

■ **cell cycle**

the necessary sequence of growth and synthetic stages through which a cell passes, from its origin by mitosis of a parent cell until its own division into daughter cells; consists of a G1 (first gap) period, an S (synthesis) phase, a G2 period and an M (mitosis) period.

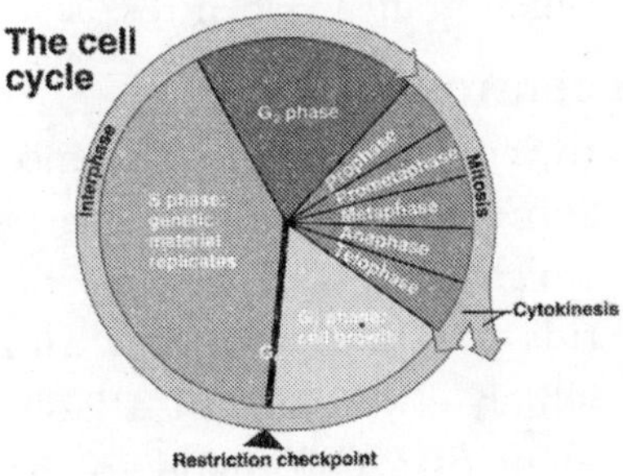

■ **cell line**

an established clone originally derived from a whole organism through a long process of cultivation.

■ **cell lineage**

the pedigree of cells resulting from binary fission.

■ **cell wall**

a tough outer coating found in many plant, fungal, and bacterial cells that accounts for their ability to withstand mechanical stress or abrupt changes in osmotic pressure. Cell walls always contain a carbohydrate component and frequently also a peptide and a lipid component.

Plant Cell Wall

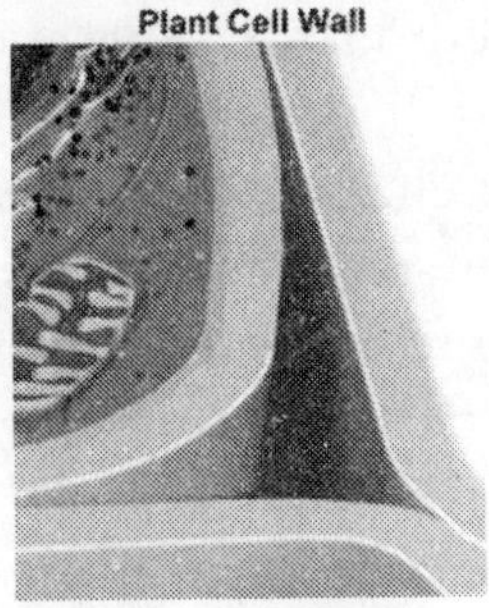

■ **centimorgan (cM)**

a genetic measure of the distance that separates markers in the same chromosome, especially in describing a map of the chromosome; the distance that permits a 1% frequency of crossing over (1 morgan is the distance that permits 100% crossing over); equivalent to about 1310^6 bases.

■ **central dogma**

in biology, the proposition that the permanent repository

of genetic information is DNA which can be replicated, and that the information is expressed unidirectionally by transcription into RNA and thence by translation into protein; later qualified in important ways, e.g. upon the discovery of reverse transcription.

■ **centrifugal elutriation**

a method for separation of isolated cells according to their characteristic sedimentation rates in a centrifuge rotor which is designed to allow flow-through of a fluid during operation; also known as countercurrent elutriation and elutriation centrifugation. The centrifugal force on the cells is opposed by the force of a fluid moving in the opposite direction. Cells are first sedimented in a density gradient, then displaced by a buffer of increasing density that flows into the bottom of the sample cell and out of the top, and carries with it cells of the same density.

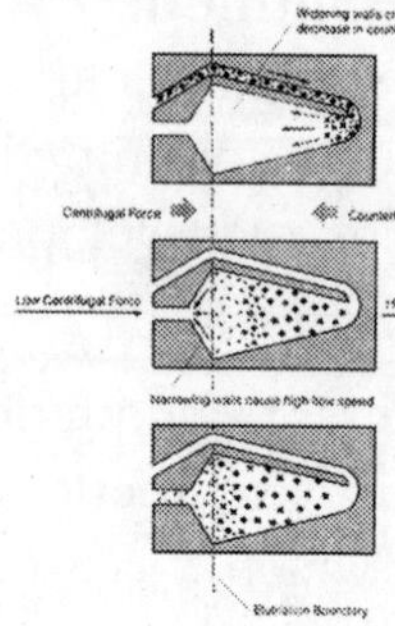

■ **centriole**

one of two cylindrical structures that appear in animal cells. During mitosis the centrioles are the poles from which the microtubules of the mitotic spindle radiate; they represent specialised microtubule-organising centres.

■ **centromere**

the constriction of a chromosome seen in metaphase at which pairs of sister chromatids are held together and which is the site of attachment of a chromatid to the mitotic spindle.

■ **centrosome**

one of the structures of a cell that during mitosis serves as a microtubule-organising centre; it becomes the structure from which the mitotic spindle radiates and which defines a cell pole.

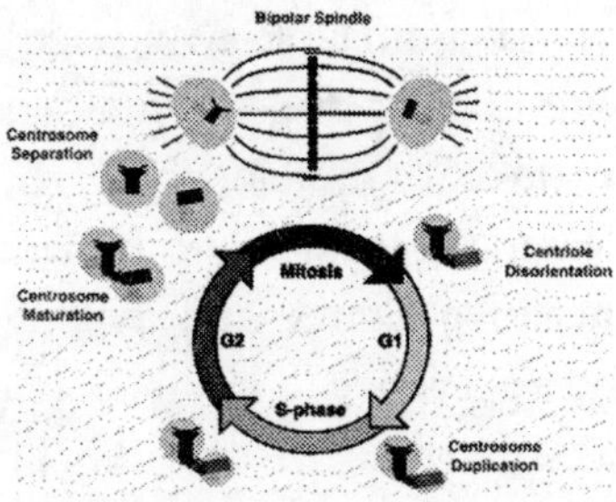

■ **cerebroside**
sphingolipid containing one sugar residue as a head group, e.g. glucosylceramide.

■ **chair form**
a conformation of a six-membered non-aromatic ring that places all opposing centres away from each other, as contrasted with the boat form, which places two of the opposing centres towards each other. A twist conformation is a slightly flattened and relaxed variation of the chair form.

■ **chalone**
an inhibitory hormone.

■ **channel former**
a structure in a membrane that, without itself moving, allows passive diffusion of ions across the membrane.

■ **channeling**
the direct transfer of a reaction product from the active site of one enzyme directly to the active site of a different enzyme, where it serves as a substrate.

■ **chaotropic agent**
a solute that disrupts the structure of the bulk water phase and, in so doing, changes the solubility and stability properties of other solutes, such as proteins.

■ **chaperone machine**
a multicomponent system that ensures the proper folding of nascent proteins or their intracellular transport following synthesis; components include chaperones (eukaryotes) or chaperonins (from prokaryotes, mitochondria and plasmids), which bind to unfolded proteins; also peptidyl proline isomerases and heat-shock proteins, which can serve various ancillary functions.

■ **Chargaff's rule**
an empirical finding that in DNA the frequency of A equals the frequency of T, and the frequency of G equals the frequency of C; later given a theoretical basis by the

Watson-Crick double-helix model of DNA.

■ charge

in chemistry, the integral sum of positive and negative particles that make up a species, a cation if the charge is positive, an anion, if the charge is negative. In biochemistry, the loading of a binding site with its intended ligand, e.g. the appropriate amino acid bound to its tRNA; in cell biology, essentially as in biochemistry, but on a macromolecular level, e.g. the charging of a liposome with material for transport cross a cell membrane.

■ charge ladder

a series of elution peaks seen upon capillary electrophoresis of modified versions of a protein that vary only in their electrostatic charge.

■ charge relay system

a tautomeric form of a protein in which a formal charge on one dissociating group is moved to another in the same protein. In serine proteinases, for example, an internal - (sometimes a gamma-beta) carboxylate is partially protonated by bridging via an imidazole to an active-site nucleophile, often a serine hydroxy or cysteine thiol group, leaving the nucleophile with a negative charge and making it more reactive in its attack upon a substrate molecule. This group of residues is termed the catalytic triad.

■ chelate

a molecule that contains more than one binding site and frequently binds to another molecule through more than one binding site at the same time.

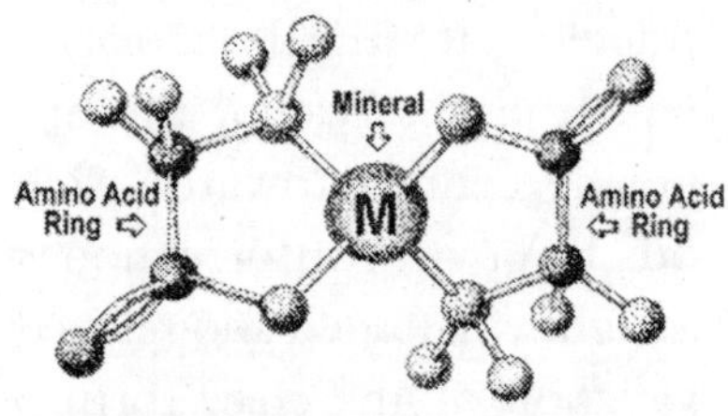

■ chelator

a compound that binds ions, especially metal ions, by several functional groups whose combined effect results in a high-affinity interaction.

■ chemical coupling hypothesis

explanation for oxidative phosphorylation by analogy

with substrate-level phosphorylation, i.e. when a phosphate derivative of an electron carrier is oxidised, it becomes a high-energy phosphate compound, energetically capable of transferring its phosphate to ADP.

■ chemical force microscopy

a variant of atomic force microscopy in which the sensing probe is coated with a substance that interacts with the test material or object.

■ chemiosmotic coupling

coupling of ATP synthesis to electron transfer *via* an electrochemical pH gradient across a membrane (inner mitochondrial membrane, bacterial plasma membrane, or thylakoid membrane).

■ chemokine

a chomotactic cytokine.

■ chemotaxis

a cell's sensing of and movement toward, or away from, a specific chemical agent.

■ chemotaxis

the movement of a cell along the concentration gradient of a chemical, the chemotactic agent, toward its source, in the case of the chemical being a chemoattractant, or away, in the case of a chemorepellant.

■ chemotroph

an organism that derives its energy from chemical reactions, usually by the oxidation of nutrients by molecular oxygen.

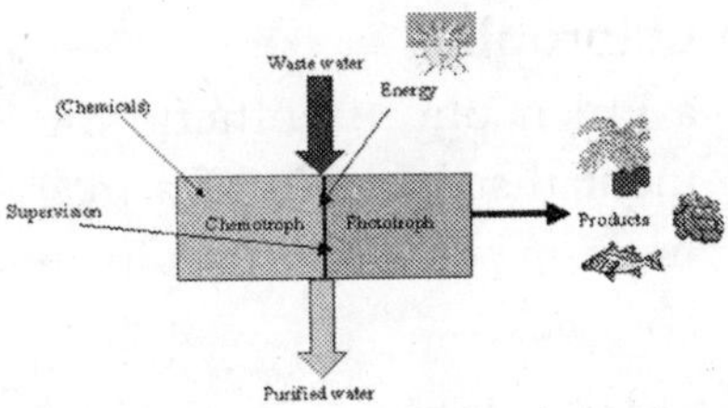

■ chemotype

a strain of bacteria defined by variations in a biosynthetic product, e.g. Gram-negative bacteria chemotypes defined by the chemistry of their lipopolysaccharides.

■ chemzyme

a chemical enzyme; a soluble small organic molecule, sometimes with the addition of a transition metal ion, that is capable of catalysing specific reactions.

■ **chimaeric DNA**

a hybrid molecule produced by combining DNA from two different species into a single polynucleotide.

■ **chiral compound**

a compound that contains an asymmetric (chiral) center and thus can occur in two nonsuperimposable mirror-image forms (enantiomers).

■ **chlorophyll**

a green photosynthetic pigment that is made of a magnesium dihydroporphyrin complex.

■ **chloroplast**

a chlorophyll-containing photosynthetic organelle, found in eukaryotic cells, that can harness light energy.

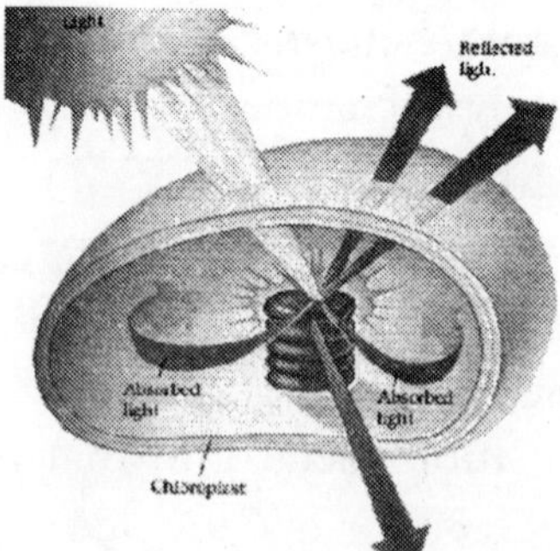

■ **Chou-Fasman analysis**

a method for prediction of the secondary structure of a protein from its primary structure. A survey of proteins of known secondary and tertiary structure allows an empirical classification of amino acid residues according to their abilities to strengthen or destabilise alpha-helix or beta-structure. Recent advances have high-speed computers using Monte Carlo calculations to first fit multi-helix proteins into individual -helices, then use empirically assigned interaction coefficients to locate the helices in spatial relation to each other.

■ **chromatin**

a filamentous complex of DNA, histones, and other proteins, constituting the eukaryotic chromosome.

■ **chromatography**

a process in which mixtures of molecules are separated by many repeated partitionings between a mobile phase and a stationary phase. The major types are size exclusion, ion exchange and affinity; can be conducted at high or low pressure.

■ **chromosome**

a single large DNA molecule and its associated proteins. The function is to store information in genes and transmit genetic information.

■ **chromosome puff**

a swollen region of a giant chromosome; the swelling reflects a high degree of transcription activity.

Puff in Polytene Chromosome

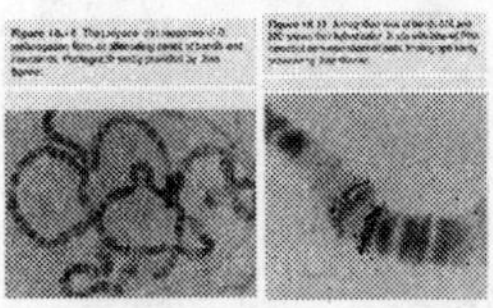

The interphase nuclei of larvae of Dipteran flies that are greatly enlarged in diameter (polytene, successive replication to 1024C)

The active sites of polytene chromosome can be visualized --> expanded as puffed state. --> transcription is demonstrated by RNA synthesis.
Both lampbrush chromosome and polytene puff indicate that genetic material is dispersed from its usual more tightly packed state.

■ **chylomicron**

a plasma lipoprotein composed of a triacylglycerol core and a shell of protein and phospholipid. The chylomicron carries lipid from the intestine to the tissues.

■ **cide**

kill.

■ **circum-**

around.

■ **cis and trans isomers**

isomers related by rotation about a double bond.

■ **cis dominance**

property of a sequence or a gene that exerts a dominant effect on a gene to which it is linked.

■ **cistron**

a unit of DNA or RNA corresponding to one gene.

■ **citric acid cycle**

Krebs cycle.

■ **clones**

the descendants of a single cell. Classically, this has meant cells, but Dolly is a clone.

■ **cloning**

the production of large numbers of identical DNA molecules or cells (or organisms) from a single ancestral DNA molecule, nucleus or cell.

■ **cloning vector**

a self-replicating entity to which foreign DNA can be covalently attached for purposes of amplification in host cells.

■ **closed system**

a system that exchanges neither matter nor energy with the surroundings.

■ **coactivator**

a molecule that functions together with a protein apoactivator. For example, cAMP is a coactivator of the CAP protein.

■ **codon**

a sequence of three adjacent nucleotides in a nucleic acid that codes for a specific amino acid.

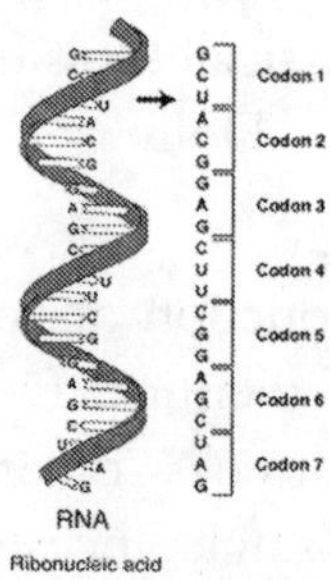

■ **coenzyme**

an organic cofactor required for the action of certain enzymes; often contains a vitamin as a component. A very common example is NADH.

■ **coenzyme A**

a coenzyme serving as an acyl group carrier in certain enzymatic reactions. Composed of pantothenic acid and adenine.

■ **coenzyme B_{12}**

an enzymatic cofactor derived from the vitamin cobalamin, involved in alkyl group transfers. Structure.

Coenzyme B_{12}

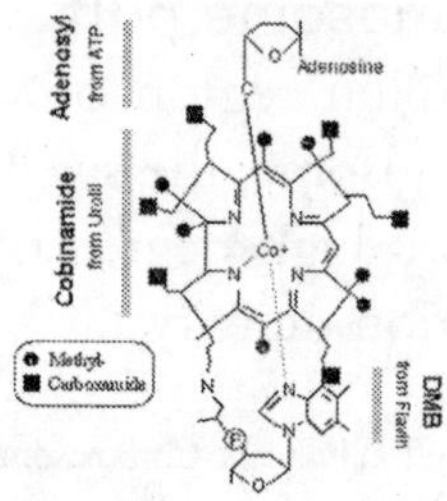

■ **cofactor**

a small molecule required for enzyme activity. It could be organic in nature, like a coenzyme, or inorganic in nature, like a metallic cation.

■ **cognate**

describing two biomolecules that normally interact; for example, an enzyme and its normal substrate, or a receptor and its normal ligand.

■ **cohesive ends**

two DNA ends in the same DNA molecule, or in different molecules, with short overhanging single-stranded segments that are complementary to one another, facilitating ligation of the ends; also known as sticky ends. Usually generated by the action of restriction enzymes.

■ cointegrate
an intermediate in the migration of certain DNA transposons in which the donor DNA and target DNA are covalently attached.

■ colligative properties
properties of solutions that depend on the number of solute particles per unit volume; for example, freezing-point depression.

■ common intermediate
referring to a metabolite; A chemical compound common to two enzymatic reactions, as a product of one and a substrate in the other.

■ competitive inhibition
a type of enzyme inhibition characterised by the inhibitor binding to the active site. The inhibition can be reversed by increasing the concentration of substrate. A competitive inhibitor is usually a structural analog of the substrate.

■ complementary
having a molecular surface with chemical groups arranged to interact specifically with chemical groups on another molecule.

■ complementary base sequence
for a given sequence of nucleic acids, the nucleic acids that are related to them by the rules of base pairing.

■ con-
together.

■ configuration
the spatial arrangement of an organic molecule that is conferred by the presence of either a double bond, about which there is no freedom of rotation, or chiral centers, around which 4 different substituent groups are arranged in a stereospecific fashion. The arrangement in either of these cases cannot be changed without breaking at least one covalent bond.

■ conformation
the spatial arrangement of substituent groups that are free to assume different positions in space without break-

ing any bonds, because of the freedom of single bond to rotate.

■ conjugate acid-base pair
a proton donor and its corresponding deprotonated species; for example, acetic acid (donor) and acetate (acceptor).

■ conjugate redox pair
an electron donor and its corresponding electron acceptor form; for example, NADH (donor) and NAD^+ (acceptor).

■ conjugated protein
a protein containing one or more prosthetic groups.

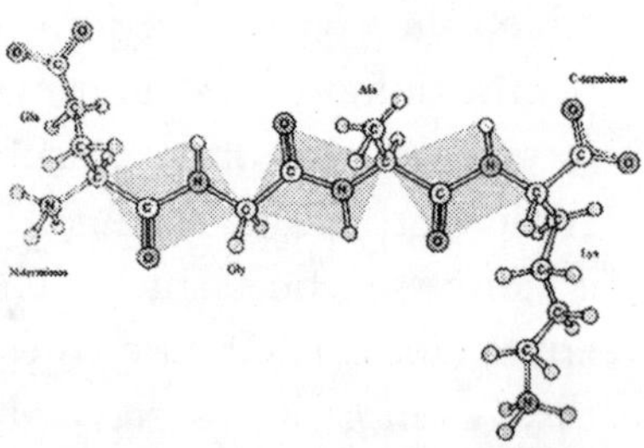

■ consensus sequence
in nucleic acids, the "average" sequence that signals a certain type of action by a specific protein. The sequences actually observed usually vary around this average.

■ conservative substitution
replacement of an amino acid residue in a polypeptide by another residue with similar properties. For example, substitution of Lys by Arg.

■ constitutive enzymes
enzymes required at all times by a cell and present at some constant level; for example, many enzymes of the central metabolic pathways. Sometimes called "housekeeping" enzymes.

■ constitutive enzymes
enzymes synthesised in fixed amounts, regardless of growth conditions.

■ contra-
against.

■ cooperative binding
a situation in which the binding of one ligand to a macromolecule favours the binding of another. For example, DNA cooperatively binds histone molecules, and haemoglobin cooperatively binds oxygen molecules.

■ **coordinate induction**
the simultaneous expression of two or more genes.

■ **Cori cycle**
the transport of the precursor and product of glycolysis between exercising muscle and the liver, i.e. lactic acid from muscle to liver, and glucose from liver to muscle.

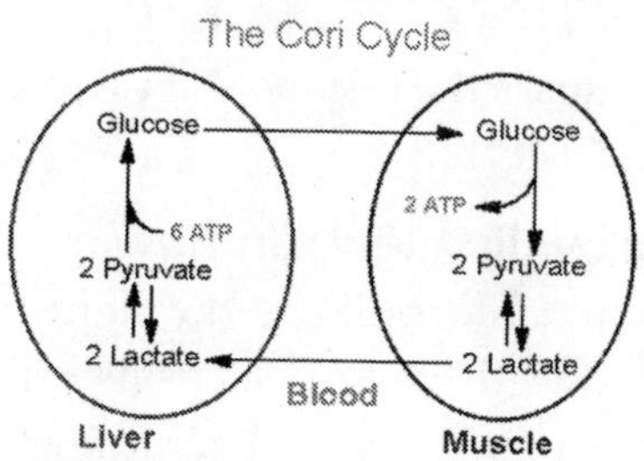

■ **Cori ester**
obsolete name for a-glucose 1-phosphate.

■ **Cornish-Bowden plot**
a graphical method for determination of the type of enzyme inhibition and the dissociation constant for an enzyme-inhibitor complex (Ki) or for an enzyme-inhibitor-substrate complex (Ki'). The effect on the enzymic rate (v) is determined at two or more substrate concentrations (S) and over a range of inhibitor concentrations (I). In a plot of S/v versus I, the data for each substrate concentration fall on straight lines that intersect at I=-Ki and S/v=Km/Vmax (uncompetitive inhibition), or intersect on the abscissa (S/v=0) at I=-Ki (non-competitive inhibition). For competitive inhibition, the lines are parallel.

■ **corticosteroids**
steroid hormones formed by the adrenal cortex.

■ **cosmid**
a cloning vector, used for cloning large DNA fragments; generally contains segments derived from bacteriophages and various plasmids.

■ **cot curve**
a curve that indicates the rate of DNA-DNA annealing as a function of DNA concentration and time.

■ **cotransport**
the simultaneous transport, by a single transporter, of two solutes across a membrane. Is either antiport or symport, depending on whether the solutes travel in opposite or the same direction.

■ **coupled reactions**
two chemical reactions that have a common intermediate and thus a means of energy transfer from one to the other.

■ **covalent bond**
a chemical bond that involves sharing of electron pairs.

■ **cristae**
infoldings of the inner mitochondrial membrane.

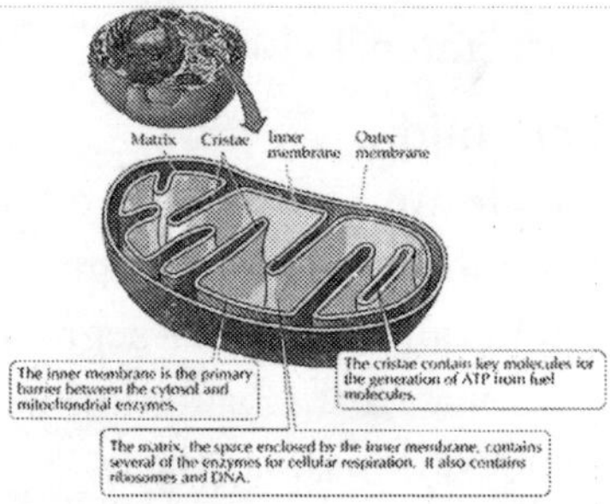

■ **c-terminal**
in a polypeptide sequence, that unique residue which is connected to the linear sequence by its amino group, leaving it with a free carboxy group. In practice, the carboxy group of a C-terminal residue may be modified, e.g. by amidation or, in the case of pyroglutamate, by internal lactamisation.

■ **c-terminus**
the carboxyl end of a protein. As proteins are linear polymers of -amino acids H_2N-CHRa-CO-(NH-CHR'-CO)$_n$-NH-CHRz-COOH where R is one of the 20 chemical groups that distinguish the amino acids that occur in proteins, the N-terminus is the amino acid residue H_2N-CHRa-CO-, and the C-terminus is the residue -NH-CHRz-COOH.

■ **cyclic electron flow**
in chloroplasts, the light-induced flow of electrons originating from and returning to photosystem I to produce ATP without production of NADPH.

■ **cyclic photophosphorylation**
ATP synthesis driven by cyclic electron flow through photosystem I (no NADPH produced).

■ **cyte-**
cell.

■ **cytidine**
a pyrimidine nucleoside found in DNA and RNA.

■ **cytochromes**

haeme proteins serving as electron carriers in cellular respiration, photosynthesis, or other redox reactions. Cytochrome c, for example.

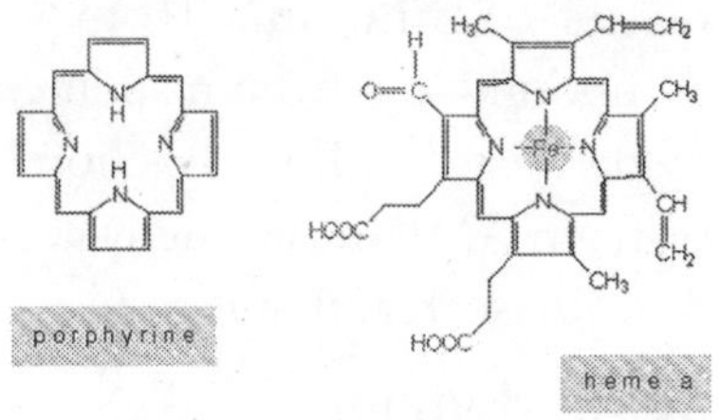

■ **cytokinesis**

the final separation of daughter cells following mitosis.

■ **cytokinin**

a plant hormone produced in root tissue.

■ **cytoplasm**

the portion of a cell's contents outside the nucleus but within the plasma membrane; includes organelles such as mitochondria, as well as soluble enzymes and solute particles.

■ **cytosine**

a pyrimidine base found in DNA and RNA.

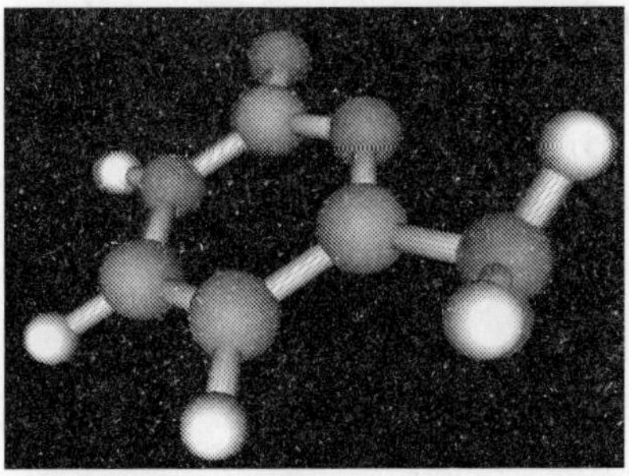

■ **cytoskeleton**

the filamentous network providing structure, organisation and motion to the cytoplasm: includes actin filaments, microtubules, and intermediate filaments.

■ **cytosol**

the aqueous phase of the cytoplasm (including dissolved solutes and enzymes, but excluding all organelles). It is most commonly the product of centrifugation at 100,000 xg for 1 h.

■ **D loop**

an extended loop of single-stranded DNA displaced from a duplex structure by an oligonucleotide.

■ **Dalton**

a unit of mass equivalent to the mass of a hydrogen atom (1.66×10^{-24} g).

■ Danielli-Davson model
a model for biological membranes; a variation of the Gortner and Grendel model in which globular proteins are tightly adsorbed to the polar groups of the exposed surfaces of the membrane.

■ dark reactions
reactions that can occur in the dark, in a process that is usually associated with light, such as the dark reactions of photosynthesis.

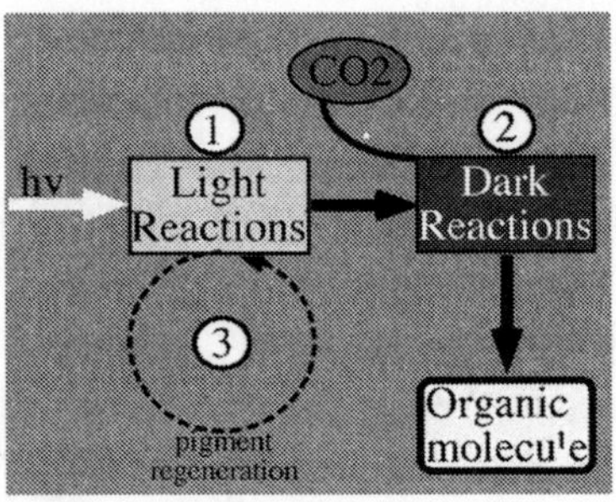

■ de-
from, away from, down from.

■ De novo pathway
a biochemical pathway that starts from elementary substrates and ends in the synthesis of a biochemical.

■ DEAD box protein
one of a group of proteins that have in common an Asp-Glu-Ala-Asp (DEAD in the one-letter code) or related sequence, and function in ATP-dependent processing of RNA.

■ deamination
the abstraction of the elements of ammonia from a compound, e.g. from histidine by the histidine lyase reaction, or from AMP in the adenylate deaminase reaction.

■ death domain
a sequence of a death-associated protein or its gene, which is common to several death-associated proteins.

■ death gene
a gene whose expression is associated with apoptosis, e.g. a Ca^{2+}-activated endonuclease that cleaves exposed regions of chromatin to produce nucleosome-sized fragments.

■ deca-
ten.

■ degenerate code
a code in which a single element in one language is specified by more than one ele-

ment in a second language. Example.

■ **dehydrogenases**
enzymes catalysing the removal of pairs of hydrogen atoms from their substrates.

■ **deletion mapping**
the assignment of the relative positions of deletion mutations along chromosomes. In the first phase, many individuals are studies for the presence of cytogenetic markers and later phases use restriction fragment length polymorphisms to detail the chemical nature of the aberrations. Futhermore, an experimental approach for identifying the location of a point mutation by testing whether recombination with a gene with a known deletion will produce the wild-type protein, in which case the site of point mutation lies outside the deleted region, or whether it fails to produce the wild-type protein, in which case the point mutation lies within the deleted region. A sufficiently large number of deletion mutants will narrow the location of the point mutation to a single nucleotide.

■ **deletion mutation**
a mutation caused by the absence of one of more nucleotides in the DNA sequence.

■ **delta restriction cloning**
a method for the use of a single primer to determine a DNA sequence longer than would normally be practical for a single sequencing attempt. After the first sequencing the bulk of the determined sequence is excised and the plasmid re-ligated, so that the primer is just upstream from the remaining undetermined sequence. The procedure may be repeated indefinitely until a large sequence is determined.

■ **denaturation**
the destruction of the ordered folding of a protein or nucleic acid that is required for its normal function. Protein denaturation often involves a change from a specific globular or fibrous con-

formation to a random coil; nucleic acid denaturation often involves the dissociation of a duplex into single strands.

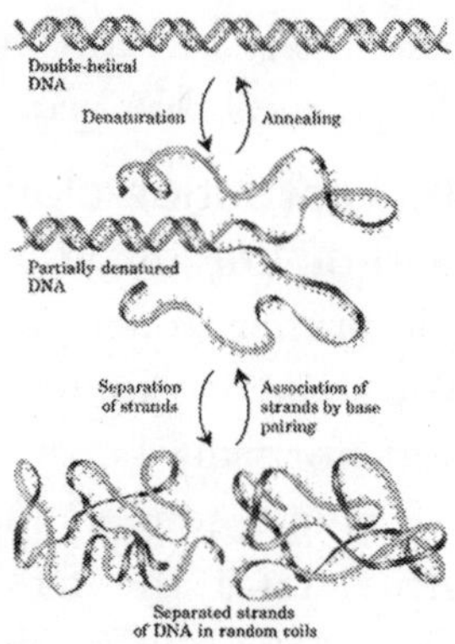

■ denaturing gradient gel electrophoresis

a method for detection of single base substitutions in DNA fragments. Putative mutant and normal double-stranded DNA fragments are applied along one edge of an agarose gel slab that contains a denaturing agent (e.g. urea and formamide) in a gradient perpendicular to the direction of electrophoresis. At a low concentration of the denaturant the fragments are more mobile, and at high concentration they are unfolded and therefore less mobile. The electrophoretic band describes a sigmoid curve on the developed gel; the denaturant concentration represents the mid-point of the curve and is characteristic of the sequence. Introduction of a GC-rich sequence, which is very stable to denaturation and thus acts as a GC clamp, into the DNA allows detection of mutations in the more stable regions of the DNA.

■ dendrogram

a phylogenetic tree.

■ density-gradient centrifugation

a technique to separate and/or characterise a macromolecule by high-speed centrifugation in a density gradient formed by the concentration gradient of a solute such as CsCl or sucrose. The macromolecule sediments until it reaches the zone of its own density. Also known as isopycnic centrifugation or zonal centrifugation.

■ **deoxyribonucleic acid (DNA)**
a polynucleotide having a defined sequence of deoxyribonucleotide units covalently bonded *via* 3',5'-phosphodiester bonds. Functions in the transmission of genetic information.

■ **deoxyribonucleotides**
nucleotides containing 2-deoxy-D-ribose as the pentose component.

■ **depside**
an ester of two or more derivatives of orsellinic acid (1,2-dihydroxy-5-methylbenzoic acid). A depsidone is a depside in which the aromatic subunits are additionally coupled to each other by an ether linkage.

■ **depurination**
the cleavage of N-glycosidic bonds of DNA to form apurinic DNA.

■ **desaturases**
enzymes (dehydrogenases) that catalyse the introduction of double bonds into the hydrocarbon portion of fatty acids.

■ **desensitisation**
the loss of responsiveness of an enzyme to allosteric regulation while retaining its catalytic activity.

■ **desmosome**
a type of junction that attaches one cell to its neighbour.

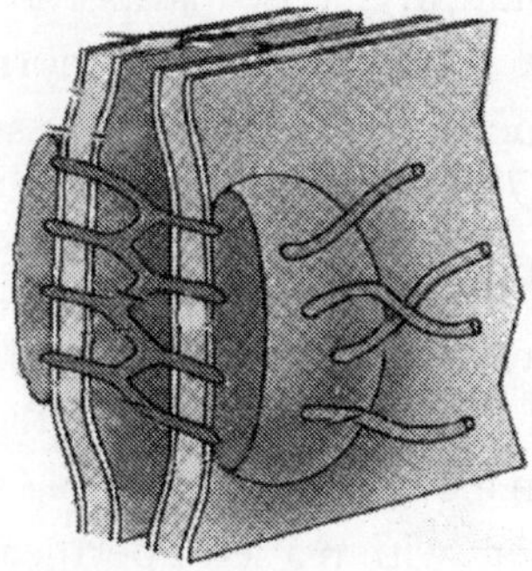

■ **desolvation**
in aqueous solution, the release of bound water surrounding a solute such as an enzyme.

■ **detoxification**
the chemical modification by oxidation, methylation, glycosylation, etc. of a xenobiotic to render it innocuous.

■ **dextrorotatory isomer**
a stereoisomer that rotates the plane of plane-polarised light clockwise.

■ di(s)-
two.

■ dia-
through, complete.

■ diabetes mellitus
a complex endocrine-metabolic disease resulting from insulin deficiency; characterised by a failure in glucose transport from the blood into cells at normal glucose concentrations, resulting in hyperglycemia.

■ dialysis
a technique for the separation of macromolecules from smaller molecules by placing them within a semi-permeable membrane, such as Cellophane, separating them from a large volume of water. Only the low-molecular-mass diffusible molecules cross the membrane and pass into the larger volume; the macromolecules are confined to their original space. Equilibrium dialysis is the technique of quantification of binding capacity and affinity by dialysis of a macromolecule against various concentrations of a ligand and subsequent measurement of the final concentrations of bound and free ligand within the dialysis chamber and free ligand outside it.

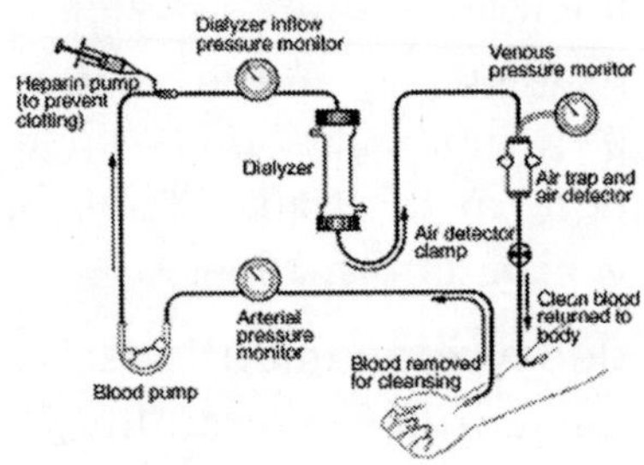

■ diaphorase
an enzyme that transfers electrons from NADH to a dye or to ferricyanide.

■ diastereoisomer
one of two or more compounds that differ from each other at one or more asymmetrical centres, e.g. D-erythrose and D-threose.

■ diauxic growth
biphasic growth on a mixture of two carbon sources in which one carbon source is used up before the other one.

■ diauxie
bacterial growth in two stages which occurs when the bacteria are grown on two carbon sources, e.g. glucose and xylose. The first

phase corresponds to consumption of one compound, followed by a lag phase until enzymes for the assimilation and metabolism of the second compound are produced.

The Diauxic Shift: Overview of Gene Expression Changes

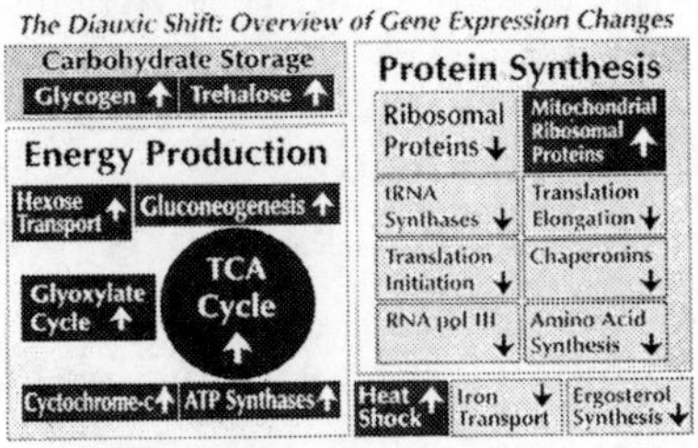

■ **difference cloning**

a group of techniques (e.g. subtractive DNA cloning) designed to isolate DNA sequences that are not shared between two DNA sources, e.g. cDNA clones prepared from different tissues, or under different conditions. Difference cloning is distinguished from coincidence cloning, comprising techniques (e.g. end ligation coincident sequence cloning) designed to isolate DNA sequences that are shared between two DNA sources.

■ **difference spectra**

plots comparing the absorption spectra of a molecule or an assembly of molecules in different states, for example, those of mitochondria under oxidising or reducing conditions.

■ **differential centrifugation**

separation of cell organelles or other particles of different density by their different rates of sedimentation in a centrifugal field.

■ **differentiation**

a change in the form and pattern of a cell and the genes it expresses as a result of growth and replication, usually during development of a multicellular organism.Also occurs in microorganisms (e.g.in sporulation).

■ **diffusion**

the net movement of molecules in the direction of lower concentration.

■ **diffusion coefficient**

a measure of a molecule's ability to travel through a solution, propelled by Brownian movement; expressed as RT/

Nf (cm^2·s^{-1}), where R is the gas constant, T is the absolute temperature, N is Avogadro's number and f is the frictional coefficient.

■ digestion

enzymatic hydrolysis of nutrients in the gastrointestinal system to yield their substituent components.

■ dimer

structure resulting from the association of two subunits.

■ dioxygenase

an enzyme that reduces molecular oxygen by incorporating both atoms into its substrate, e.g. tryptophan dioxygenase.

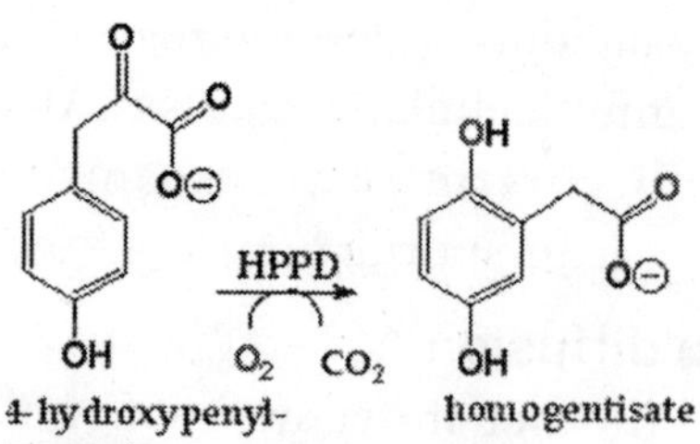

■ diplo-

double.

■ diploid

having two sets of genetic information; describing a cell with two chromosomes of each type.

■ diploid cell

a cell that contains two chromosomes (2N) of each type.

■ dipole

a molecule having both partial positive and partial negative charges.

■ diprotic acid

an acid having two dissociable protons, for example H_2SO_4 or oxalic acid.

■ direct calorimetry

evaluation of the heat evolved by a human or experimental animal by measurement of the heat exchanged with the environment in specially constructed insulated chambers.

■ directed mutagenesis

in a DNA sequence, an intentional alteration that can be genetically inherited.

■ disaccharide

a carbohydrate consisting of two covalently joined monosaccharide units. Sucrose for example, is composed of glucose and fructose.

■ **displacement chromatography**

a column chromatographic method for separation and concentration of components of a solution. After adsorption of the sample to be separated on to the resin in a column, all bound molecules are displaced by a concentrated solution of a displacer, whose affinity for the stationary phase must be higher than that of any of the sample molecules. As the displacer sweeps the adsorbed molecules ahead of itself, each of these becomes a displacer for any molecule of lower affinity. Thus the original components are eluted in small volumes ahead of the displacer, in reverse order with regard to their affinities for the stationary phase.

■ **dissociation constant**

an equilibrium constant for the dissociation of one species into two. Examples would be the dissociation of a complex of two or more biomolecules into its components, for example, dissociation of a substrate from an enzyme, descibed as a Kd; an acid dissociating into its conjugate base and a proton, described as a Ka; the dissociation of water into protons and hydroxide ions, described as a Kw; or a base dissociating into its conjugate acid and a hydroxide ion , described as a Kb.

■ **disulphide bridge**

a covalent cross link between two polypeptide chains formed by a cystine residue (two Cys residues oxidised together).

■ **Dixon plot**

a graphical method for determination of the type of enzyme inhibition and the dissociation constant (Ki) for an enzyme-inhibitor complex. The effect on the enzymic rate (v) is determined at two or more substrate concentrations, and over a range of inhibitor concentrations (I). In a plot of 1/v against I, data for each substrate concentration fall on straight lines that intersect at I=-Ki and 1/v=1/Vmax (competitive inhibition), or that intersect on the abscissa (1/v=0) at I=-Ki (non-competitive inhibition). For uncompetitive inhibition, the lines are parallel.

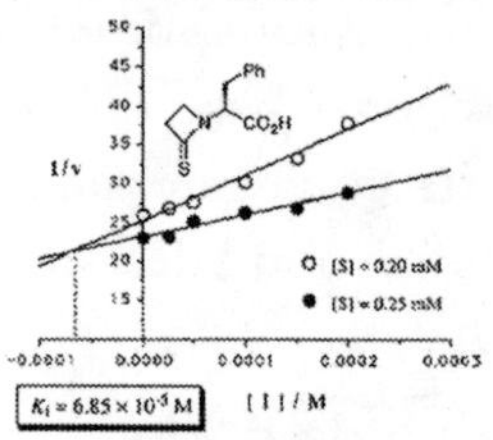

Dixon plot for the inhibition of CPA-catalyzed hydrolysis of hippuryl-L-phenylalanine by 2 at [CPA] = 1.2×10^{-6} M.

■ DNA chimera

a DNA containing genetic information derived from two different DNA molecules.

■ DNA cloning

the production of large numbers of identical DNA molecules or cells from a single ancestral DNA molecule, nucleus or cell.

■ DNA library

a random collection of cloned DNA fragments designed to include all or most of the genome of a given organism; also called a genomic library.

■ DNA ligase

an enzyme that creates a phosphodiester bond between the 3' end of one DNA segment and the 5' end of another.

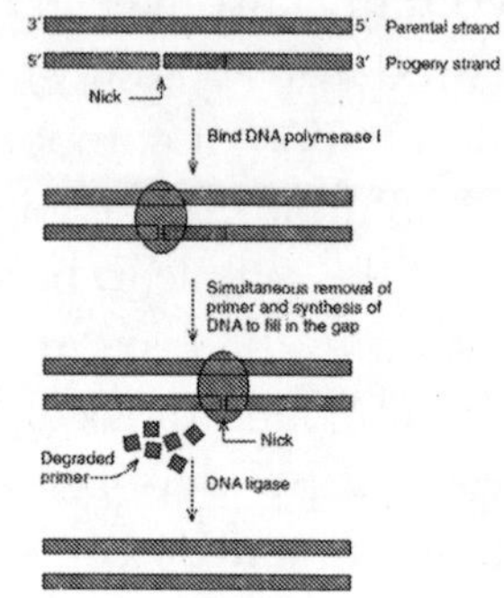

■ DNA molecular decoy

a synthetic polynucleotide with high affinity for a regulatory protein, such as a transciption factor, that can be used for research purposes, and potentially for therapy, to compete with the natural DNA sequence and attenuate the effects of the regulatory protein.

■ DNA polymerase

an enzyme that catalyses template-dependent synthesis of DNA from its dNTP precursors.

■ DNA replicase system

the entire complex of enzymes and specialised proteins required in biological DNA replication.

■ **DNA supercoiling**
the coiling of DNA upon itself, into a more condensed structure.

■ **DNA transposition**
the movement of a gene or set of genes from one site in the genome to another.

■ **docking**
in a computer graphics simulation, the bringing together of two molecular models to explore their interactions, e.g. the insertion of a substrate into the active site of an enzyme.

■ **dolicho-**
long.

■ **domain**
a segment of a folded protein structure showing conformational integrity.A domain could include the entire protein or just a fraction of the protein. Some proteins, such as antibodies, contain many structural domains.

■ **dominant**
describing an allele whose phenotype is expressed regardless of whether the organism is homozygous or heterozygous for that allele.

■ **DOP**
degenerate oligonucleotide primed PCR.

■ **dot blotting**
a method to estimate the concentration of a polynucleotide or protein solution by spotting it on a sheet of nitrocellulose or nylon, hybridising it with a radiolabelled complementary polynucleotide or antibody, visualising it by autoradiography, then comparing its intensity with that of a similarly treated graded concentration series of the standard polynucleotide or protein.

■ **double adaptor method**
cDNA concatamer sequencing; a method for increasing the efficiency of sequencing cDNA library subclones cDNA fragments are cleaved with a restriction nuclease and randomly religated into larger DNA fragments (mis-named concatamers), which are then randomly sheared and the ends, repaired. A 5'-overhanging adaptor is ligated to them, which allows their insertion into

a vector which has been prepared by leavage and religation with complementary oligonucleotide 5'- overhangs. The vectors with inserts are used for transfection. When these inserts are sequenced, the artificial junction between religated fragments may be recognised by the presence of the restriction site. The method increases the efficiency of sequencing of what would otherwise be 3'-ends of fragments, which are less likely to be successfully sequenced.

■ double helix

the natural coiled conformation of two complementary, antiparallel DNA chains by the formation of A-T and G-C base pairs.

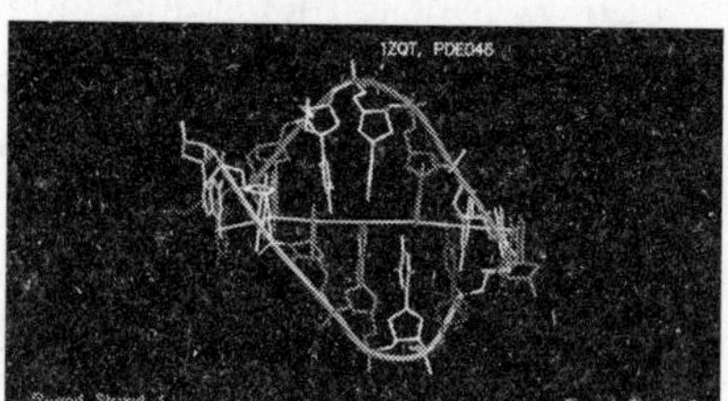

■ double minute chromosome (DMC)

an autonomously replicating structure found, for example, in leukaemic cells.

■ double sieve

descriptive of the mechanism for assuring correct charging of a tRNA by its amino acid activating enzyme. Only an amino acid that exactly fits the binding site will be ligated to the nucleic acid; those that are too large do not bind to the active site and those that are too small are hydrolysed.

■ double-reciprocal plot

a plot of 1/Vo versus 1/[S], which allows a more accurate determination of Vmax and Km than a plot of Vo versus [S]; also called the Lineweaver-Burk plot.

■ downfield

descriptive of a resonance position in a magnetic field lower than that at which a standard displays its signal.

■ down-shift

the response by cells, especially micro-organisms, to an altered nutritional state, e.g. the decrease in ribosomes or enzyme production when levels of some nutrients are low.

■ downstream

in a polynucleotide chain, towards the 3' end.

■ duplex
same as double helix.

■ dur-
hard, firm.

■ dyad symmetry
property of a structure that can be rotated by 180° to produce the same structure.

■ dys-
bad, abnormal.

■ e-, ec-
out, from out of.

■ E. coli (Escherichia coli)
a common bacterium found in the small intestine of vertebrates; the most well-studied organism.

■ Eadie-Hofstee plot
a linearised display of kinetic data of dependence of enzyme activity on substrate concentration; rate is plotted against the ratio of rate to substrate concentration.

■ Easson and Stedman plot
a graphical method for determination of enzyme-inhibitor dissociation constants for non-competitive inhibitors, especially those that bind very tightly. Where I is the total inhibitor concentration, E is the total enzyme concentration and is the ratio of inhibited to non-inhibited rates under the same conditions, I/ is plotted on the ordinate (y-axis) against 1/(I-) on the abscissa (x-axis) to give a straight line that intersects the ordinate at E; its slope is Kd, the dissociation constant.

■ ecdysone
a hormone that stimulates the molting process in insects.

■ ecto-
outside, external.

■ ectoenzyme
a plasma-membrane-associated enzyme that catalyses reactions in the extracellular space, e.g. acetylcholinesterase, 5½-nucleotidase.

■ ectomy
removal of, cut out.

■ editosome
a hypothetical multienzyme complex which in trypanosomes performs RNA editing, especially the deletion and insertion of U residues.

Edman degradation
a chemical technique to degrade and cleave amino acid residues sequentially from a protein beginning at the N-terminus, and to identify the residues as they are removed. Reaction of the N-terminal residue with phenylisothiocyanate cleaves it from the protein as the phenylthiohydantoin derivative, which may be isolated and identified.

effector
a compound that modulates an allosteric enzyme; a system that produces an intracellular response to a hormone, e.g. adenylate cyclase.

eicosanoid
one of a class of compounds that includes the prostaglandins, thromboxanes and leukotrienes, all derived from arachidonic (eicosatetraenoic) acid.

electrochemical gradient
the sum of the gradients of concentration and of electric charge of an ion across a membrane. Used as the driving force for processes such as oxidative phosphorylation and photophosphorylation.

electrochemical potential
the energy required to maintain a separation of charge and of concentration across a membrane.

electrogenic
contributing to the generation of an electrical potential across a membrane.

electromicrofiltration
a separation procedure in which an electric field is imposed upon an ultrafiltration membrane in order to prevent non-permeable material from accumulating on and clogging it.

electron acceptor
a substance that receives electrons in a redox reaction.

electron carrier
a protein or other chemical species, such as a cytochrome or coenzyme Q, that can reversibly gain and lose electrons. They function in the transfer of electrons from organic nutrients to oxygen or

some other terminal acceptor in the case of anaerobes.

■ **electron density map**

in X-ray crystallography, a representation that resembles a geological survey map of planes drawn through a crystal. It shows contours of equal electron density surrounding the individual atoms: steeper around heavier atoms; more shallow around lighter atoms.

■ **electron donor**

a substance that donates electrons in a redox reaction.

■ **electron nuclear double resonance (ENDOR)**

a technique for detection of coupling between electrons and nuclei in order to gain information about the valance electron distribution around a paramagnetic nucleus; observation of an electron spin resonance transition while applying radio frequency energy to effect nuclear magnetic resonance transitions.

■ **electron transport**

the mediation of oxidation of one metabolite and the reduction of another by a series of carriers, cytochromes, iron-sulphur proteins, quinones, etc.

■ **electron-transfer potential**

a quantitative measure of the ability of a molecule to lose an electron in a redox reaction under standard conditions; expressed in volts (V).

■ **electropherogram**

a graphical representation of an electrophoretic separation; e.g. for a slab gel, absorbance (ordinate) plotted as a function of distance from the origin (abscissa).

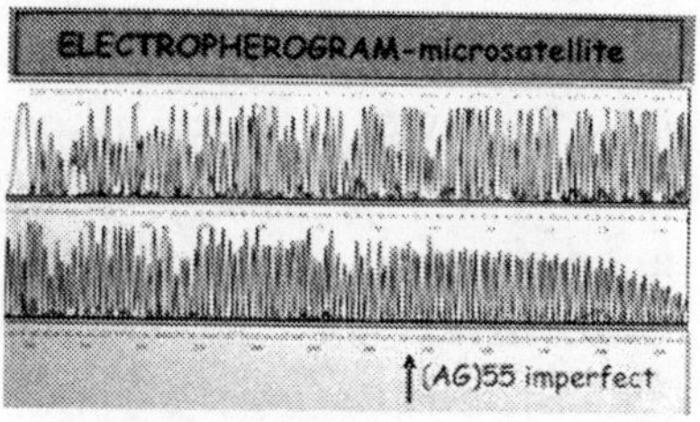

■ **electrophile**

an electron-deficient group with a strong tendency to accept electrons from an electron-rich group (nucleophile).

■ **electrophoresis**

a technique that separates charged compounds according to their mobilities in an electric field. In free zone electrophoresis the compounds are in a solution unsupported by any stabilising matrix. In gel electrophoresis the solution is stabilised by a gel (e.g. agarose, polyacrylamide, starch). In paper electrophoresis the stabilising matrix is buffer-soaked paper. In isotachophoresis, by contrast, charged solute molecules all travel at the same rate: a highly mobile leading ion (e.g. Cl-) travels first and a relatively immobile ion (e.g. tricine) is the trailing ion; to avoid an electrostatic gap, other solute molecules travel between them, all at the same rate but ordered according to their relative mobilities. Capillary eletrophoresis uses a long, very-narrow-bore tube coated with the stationary phase; as the mixture passes along the length of the lumen, components are differentially retarded to effect a separation.

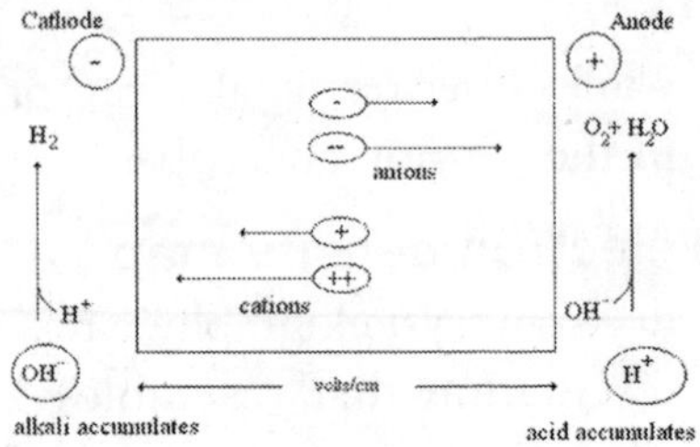

■ elongation cycle

the reactions of a ribosome that add one amino acid residue to the C-terminus of a growing polypeptide chain and move the ribosome three nucleotides towards the 3'-end of the mRNA.

■ elongation factors

protein factors uniquely required during the elongation phase of protein synthesis. Elongation factor G (EF-G) brings about the movement of the peptidyl tRNA from the A site to the P site of the ribosome.

■ eluate

the effluent from a chromatographic column.

■ em-

in.

■ embryo

plant or animal at an early stage of development.

■ en-
into.

■ enantiomers
stereoisomers that are nonsuperimposable mirror images of each other.

■ enantiomorphs
isomers that are mirror images of one another.

■ endergonic
descriptive of a chemical reaction that consumes free energy.

■ endergonic reaction
a chemical reaction that consumes energy (that is, for which Δ G is positive). These are nonspontaneous processes.

■ endo-
into.

■ endocrine
descriptive of secretion into the bloodstream, e.g. insulin secretion.

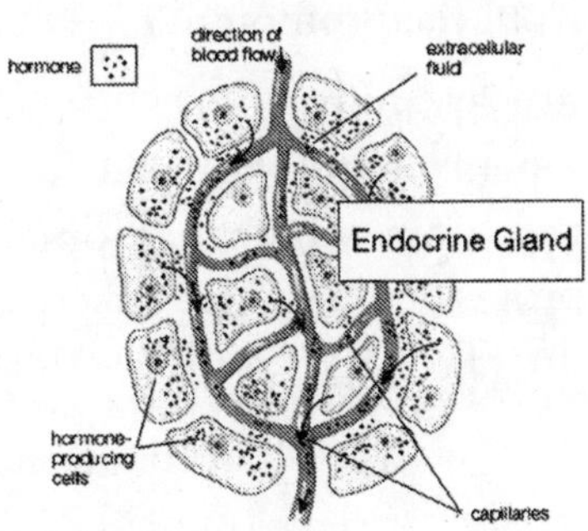

■ endocrine disrupter
an environmental chemical, e.g. an herbicide, that unintentionally mimics or interrupts the normal action of a hormone and thus causes medical or biological problems in humans or animals.

■ endocrine glands
groups of cells specialised to synthesise hormones and secrete them into the blood to regulate other types of cells. The pancreas, for example.

■ endocytosis
the uptake of extracellular material by its inclusion within a vesicle (endosome) formed by an invagination of the plasma membrane. qv receptor-mediated endocytosis.

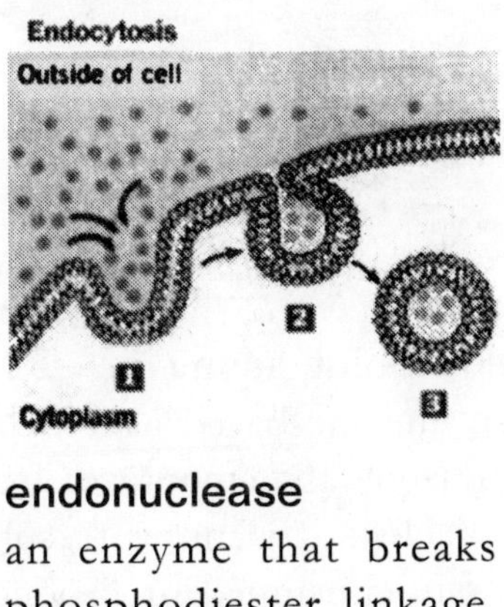

■ endonuclease
an enzyme that breaks a phosphodiester linkage at

some point within a polynucleotide chain.

■ **endopeptidase**

an enzyme that breaks a polypeptide chain at an internal peptide linkage.

■ **endoplasmic reticulum**

a system of double membranes in the cytoplasm that is involved in the synthesis of transported proteins. The rough endoplasmic reticulum has ribosomes associated with it. The smooth endoplasmic reticulum does not.

■ **endothermic reaction**

a chemical reaction that takes up heat (that is, for which ΔH is positive). These reactions are spontaneous.

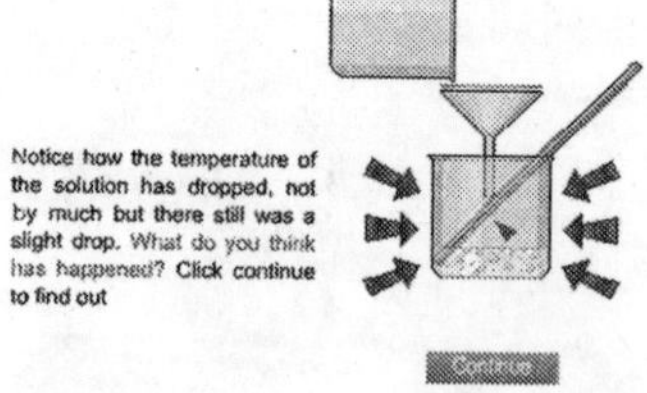

■ **end-point assay**

an enzyme-based assay that measures the amount of material by the quantity of a substrate consumed or product formed over the course of a reaction.

■ **end-product inhibition**

inhibition of an allosteric enzyme at the beginning of a metabolic sequence by the end product of the sequence.

■ **energy charge**

the fractional degree to which the ATP/ADP/AMP system is filled with high-energy phosphate groups.

■ **energy coupling**

the transfer of energy from one process to another. For example, the coupling of ATP hydrolysis with glucose phosphorylation.

■ **enhancer**

a DNA sequence that can stimulate transcription at an appreciable distance from the site where it is located. It acts in either orientation and either upstream or downstream from the promoter.

■ **enkephalin**

a pentapeptide isolated from the brain which has opiate properties, e.g. Met-enkephalin (Tyr-Gly-Gly-Phe-Met),

Leu-enkephalin (Tyr-Gly-Gly-Phe-Leu).

Example of a cyclic prodrug of Leu-enkephalin

Cyclic prodrugs are synthesized in order to increase the membrane permeability of peptides, e. g. Leu-enkephalin.

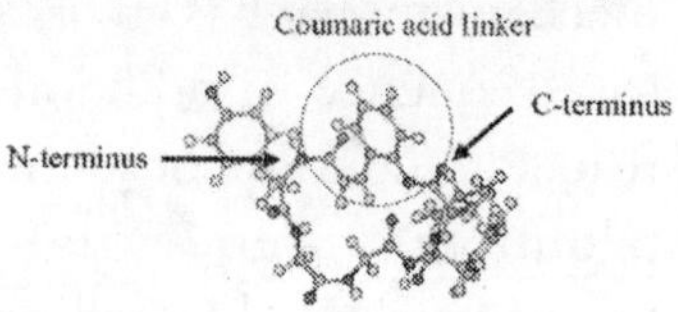

■ **ent-**

within.

■ **enterohepatic circulation**

the movement of a metabolite, e.g. a bile acid, from secretion by the gall bladder into the lumen of the intestine, its absorption, transport to the liver, and re-secretion in the bile.

■ **enthalpy**

a measure of the internal energy of a system, comprising binding forces, pressure, etc., expressed as J/mol (or cal/mol).

■ **enthalpy change (ΔH)**

for a reaction, is approximately equal to the difference between the energy used to break bonds and the energy gained by the formation of new ones.

■ **entropy**

a measure of the disorder or randomness of a system, expressed in J/mol per degree K (or cal/mol per degree K).

■ **entropy effect**

the acceleration of a reaction that occurs when reactive groups are constrained in a productive orientation, either intermolecularly as on an enzyme surface, or intramolecularly as in a model compound. This phenomenon has been referred to variously as antichimaeric assistance, approximation, catalytic configuration, the Circe effect, coupling between conformational fluctuations, directed proton transfer, electric field effect, electrostatic stabilisation, FARCE (freezing at the reactive centres of enzymes), gas-phase analogy, group transfer hydration, orbital perturbation theory, orbital steering, propinquity effect, proximity effect, rotamer distribution, stereoelectronic control; stereopopulation control, substrate anchoring, togetherness, torsional stress and vibrational activation.

envelope conformation
a conformation of a five-membered ring, e.g. a furanose, in which four ring atoms lie in a plane and C-2 or C-3 (2-endo or 3-endo) is out of the plane.

enzyme
one of a class of biological catalysts that is composed principally of a globular protein of one or more polypeptide chains. In some cases enzymes include covalently bound or tightly associated metal ions, prosthetic groups or carbohydrates; they range in molecular mass from around 10000 to several hundred thousand Da.

enzyme mechanism
an elaborate system of nomenclature has been proposed to describe, from the perspective of kinetics, the variety of enzyme mechanisms. Uni, Bi, Ter and Quad refer to the number of substrates or products (e.g. Bi Bi refers to an enzyme that is bi-reactant in both directions, and uni bi to an enzyme that is uni-reactant in the forward direction and bi-reactant in the reverse direction). Ordered describes a mechanism that has an obligatory sequence of addition of reactants (and dissociation of products), sequential a mechanism in which all reactants are added before any product is released, and random a mechanism where there is no obligatory sequence. Ping Pong describes cases in which one or more products are released before all of the substrates have added to the enzyme. Iso refers to situations in which the enzyme undergoes isomerisation to another stable form during the catalytic cycle.

E_0
the electromotive force exhibited at an electrode by 1 M concentrations of a reducing agent and its oxidised form at 25 °C and pH 7.0; a measure of the relative tendency of the reducing agent to lose electrons.

■ **eosinophil**

a polymorphonuclear leucocyte containing granules that react with eosin, a histological stain for acidic substances.

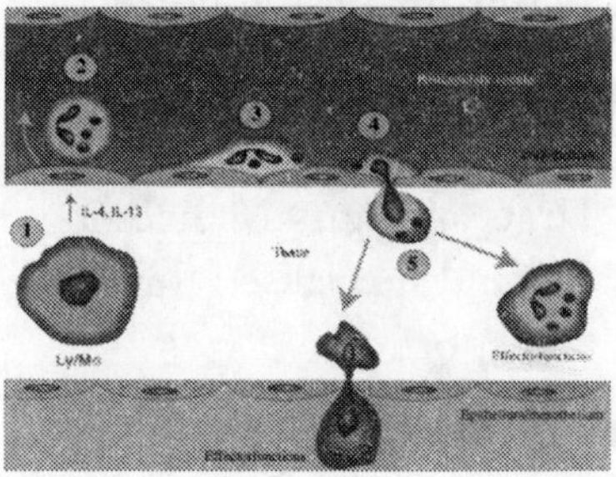

■ **epi-**

on, up, against, high.

■ **epigenetic**

descriptive of heritable changes in gene expression due to alterations in the structure of chromatin and/or methylation of DNA; e.g. genetic imprinting. Because these changes do not affect DNA sequences, they are, potentially, reversible.

■ **epigenetic state**

the condition of a cell that has been programmed during early embryogenesis, e.g. into ectoderm, endoderm or mesoderm, to a developmental fate that will be expressed many generations later.

■ **epimer**

a compound that differs from another by its configuration at only one asymmetrical centre.

■ **epimerases**

enzymes that catalyse the reversible interconversion of two epimers.

■ **episome**

a polynucleotide that can be transferred from one bacterium to another, either as a discrete structure or as a sequence incorporated into the host genome.

HPV Genome: Episome

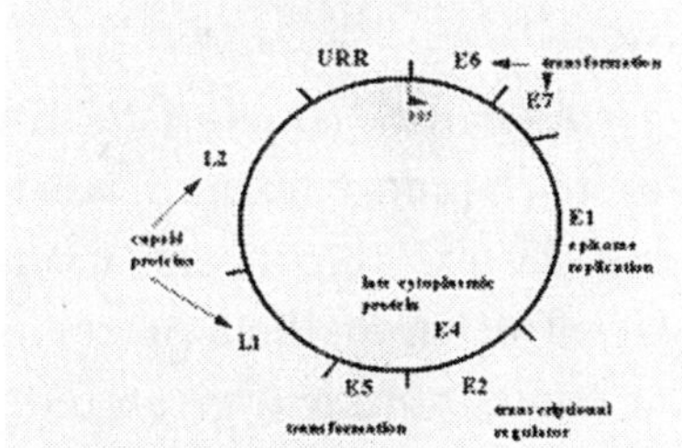

■ **epithelial cell**

any cell that forms part of the outer covering of an organism or organ.

■ **epithelium**

a sheet-like tissue that lines body vessels, cavities and surfaces, and can differentiate into secretory glands.

■ epitope

the antigenic determinant; i.e. that part of an antigen to which an antibody is directed; it defines the specificity of the antigen and consequently binds to the receptor on a T-cell or to the antibody.

■ equilibrium

the point at which the concentrations of two compounds are such that the interconversion of one compound into the other compound does not result in any change in free energy.

■ equilibrium constant (Keq')

a constant, characteristic for each chemical reaction relating the molar concentrations of all reactants and products at equilibrium at a given temperature, pressure and pH.

■ equilibrium labelling

in metabolic studies, the exposure of a biological system (e.g. a tissue slice, a cell culture, mitochondria) to a radiolabelled metabolite over a relatively long time so as to establish a constant steady-state specific radioactivity in downstream metabolites, in order to observe the metabolic fate of individual atoms of the original metabolite.

■ equine

derived from the horse.

■ erythrocyte

a red blood cell containing large amounts of haemoglobin and specialised for oxygen transport.

■ erythrocyte

an enucleated, mature, red blood cell.

■ Escherichia coli (E. coli)

a Gram negative bacterium commonly found in the vertebrate intestine. It is the bacterium most frequently used in the study of biochemistry and genetics.

■ essential amino acids

amino acids that cannot be synthesised by humans (or other vertebrates) and must be obtained from the diet.

■ essential fatty acid

a fatty acid that is a dietary necessity, e.g. the polyunsaturated fatty acids linoleic acid and linolenic acid.

■ established cell line

a group of cultured cells derived from a single origin and capable of stable growth for many generations.

■ esterase

an enzyme that hydrolyses esters; unless otherwise indicated, esters of carboxylic acids. The category includes the lipases, which, most generally defined, hydrolyse lipids; however, many lipases are active only when the lipid is in micellar form; thus they are characteristically inactive against low concentrations of a lipid, and show activity (surface activation) only when the lipid concentration rises above the critical micellar concentration.

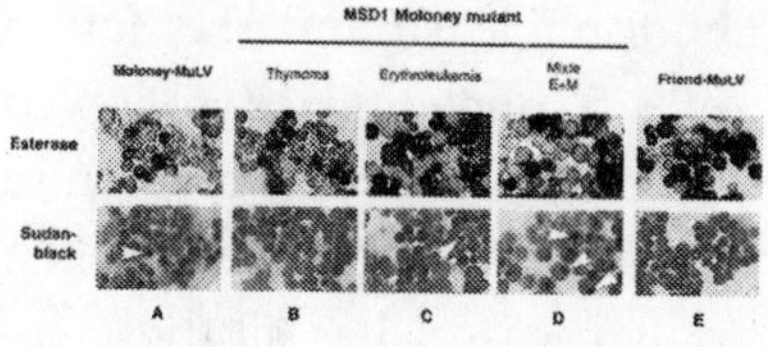

■ ether

a molecule containing two carbons linked by an oxygen atom.

■ etioplast

an organelle of a higher plant that has been grown in the dark; it contains no chlorophyll but has chlorophyll precursors and in light can develop into chloroplast.

■ eu-

true.

■ euchromatin

in cytology, the lightly staining regions of chromosomes that contain less-condensed chromatin and are the regions that are transcribed as RNA.

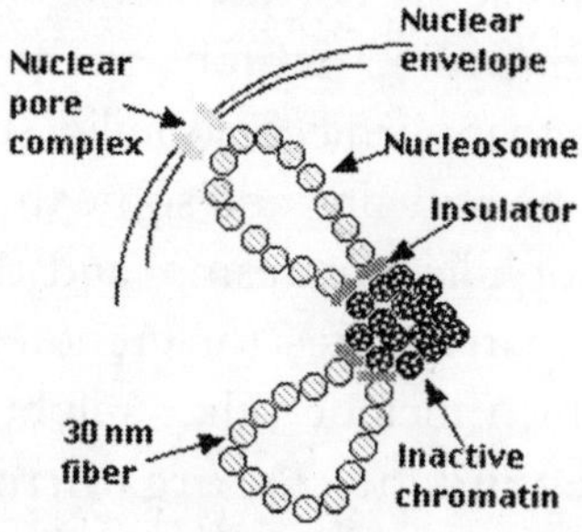

■ eukaryote

common name of a member of the Eucarya, one of the three domains of living organisms that are classified according to rRNA sequence homologies; they are further characterised by cells containing nuclei and membranes composed primarily of diacylglycerol derivatives. The other two domains are the Eubacteria (the bacteria), which have no nuclei (i.e. prokaryotes) and also have diacylglycerol derivative membrane components, and the Archaea (archaebacteria), which are also prokaryotes; they have a membrane composed of isoprenyl glycerol diether and tetraether lipids. Kingdoms within the Eucarya include plants, animals and fungi; kingdoms of the Archae are euryarchaeotes (or euryotes), which include methanogenic, halophilic, sulphur-reducing and some thermophilic organisms, and the procrenarchaeotes (or crenotes), which include other sulphur-dependent organisms, thermoacidophiles and extreme thermophiles (eocytes).

■ **euploidy**

the state of having the normal number or chromosomes, as contrasted with aneuploidy, having an abnormal number of chromosomes.

■ **ex-, exo-**

out, from out of.

■ **exaptation**

the adaptation over evolutionary time of a protein, and its gene, to a novel function; e.g. the transformation of trypsinogen into an antifreeze glycoprotein.

■ **excimer**

in fluorescence spectroscopy, a dimer of a fluorophore that is stable only in the excited state, e.g. of aromatics such as pyrene or 2,5-diphenyloxazole (PPO, used for scintillation counting).

■ **excision**

the removal of a section of double-stranded DNA that is faulty due to mutation or incorrect replication. The process includes incision, i.e. cleavage of the strand by an endonuclease and the action of a 5'-nucleotidase, followed by filling in by a polymerase and ligation. The E. coli DNA polymerase I has all these activities.

■ **excision repair**

the removal of a segment of a DNA duplex that contains a thymine or other dimer, formed by ultraviolet-light-induced damage, and resynthesis and ligation of the excised sequence.

■ excitable membrane

a membrane containing a receptor which transduces an external signal, e.g. a membrane containing acetylcholine receptors or rhodopsin.

■ excitation spectrum

in fluorescence spectroscopy, emission at a longer, fixed, wavelength as a function of the wavelength of the exciting light.

■ excited state

an energy-rich state of an atom or molecule. An electron absorbs electromagnetic energy to be promoted from the ground state to an excited state.

■ excitotoxin

a molecular signal that acts via a cell-surface receptor and initiates a series of events that result in cell death; e.g. glutamate, some of whose receptors initiate apoptotic or necrotic cascades.

■ exergonic reaction

a chemical reaction that proceeds with the release of free energy (that is, for which ?G is negative and the reaction spontaneous).

■ exocrine

descriptive of secretion into ducts that open on to body surfaces rather than into the bloodstream, e.g. salivary and gastric secretions.

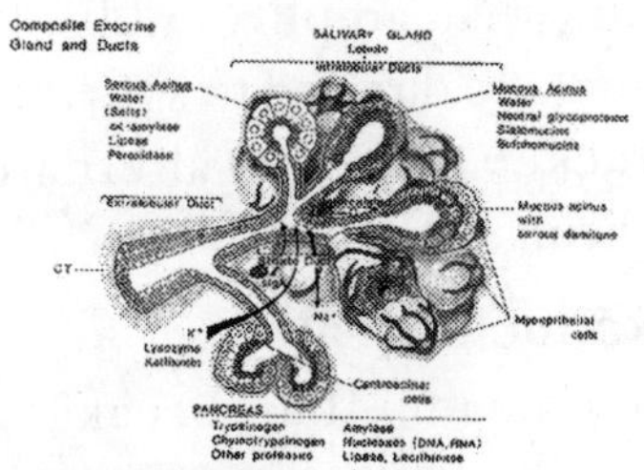

Composite diagram of exocrine glands and ducts that drain them.

■ exocytosis

a secretory mechanism whereby secretory vesicles fuse with the cell membrane and expel their contents to the exterior.

■ exogenous

arising from a source outside the organism or cell.

■ exon

the segment of a eukaryotic gene that is transcribed into a protein or incorporated into the structure of an RNA. See also intron.

■ **exon scanning**

a procedure for approximating sites of alternate splicing. In the PCR amplification of cDNA, a single primer matched to a sequence near one end, 3' or 5', of the coding sequence, is paired with primers matched to sites increasingly remote from it. The pairs that amplify multiple bands are directed to sites that bracket a site of alternate splicing.

■ **exonuclease**

an enzyme that breaks a phosphodiester linkage at one or the other end of a polynucleotide chain so as to release single or small nucleotide residues.

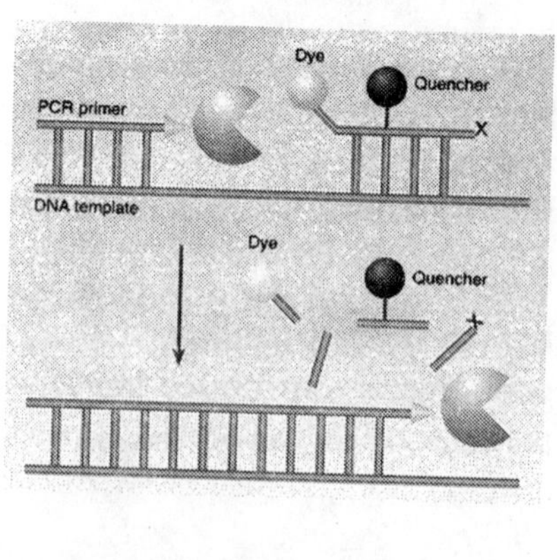

■ **exopeptidase**

a peptidase that cleaves a protein sequentially, starting at the N-terminal peptide bond (an aminopeptidase) or at the C-terminal peptide bond.

■ **exosome**

a multi-exoribonuclease complex of prokaryotic and eukaryotic cells.

■ **exothermic reaction**

a chemical reaction that releases heat (that is, for which Δ H is negative).

■ **exotoxin**

a disease-causing agent that is produced and secreted by a pathogen; contrasted with an endotoxin, which is an intrinsic component of the pathogen, e.g. the lipopolysaccharide of Gram-negative bacteria.

■ **expression cloning**

as part of a repetitive process for screening a library, the sampling of cells according to the presence of the product of the gene of interest, e.g. the identification of the glutamate receptor in a clone of cells as evidence that the receptor gene is contained within the cells.

■ **extra-**
outside, beyond, in addition.

■ **extrinsic pathway**
the blood clotting cascade that is initiated by a tissue factor.

■ **F factor**
a large bacterial plasmid, known as the sex-factor plasmid because it permits mating between F+ and F- bacteria.

■ **fab fragment**
a product of papain hydrolysis of an immunoglobulin G consisting of the variable region and some of the constant region of a heavy chain and an entire light chain, and containing a single antigen-binding site.

Fab fragment of monoclonal antibody

■ **facilitated diffusion**
diffusion of a molecule across a biological membrane *via* a protein transporter down a concentration gradient. Also called passive diffusion.

■ **factitious protein**
a product of genetic engineering; a protein designed for a specific purpose or for its expected properties.

■ **facultative aerobe**
an organism that can use molecular oxygen in its metabolism but that also can live anaerobically.

■ **facultative cells**
cells that can live in the presence or absence of oxygen.

■ **FAD (flavin adenine dinucleotide)**
the coenzyme of some redox enzymes; it contains riboflavin. Structure

■ **fat**
usually neutral fat; triacylglycerols.

■ **fatty acid**
a carboxy group on an alkyl chain that is usually unbranched; may be a short-chain (alkyl group contains

less than 7 carbons), medium-chain or long-chain (alkyl group contains more than about 11 carbons) fatty acid.

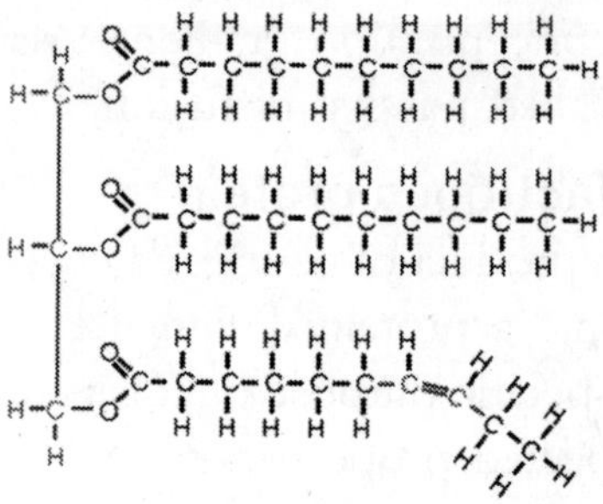

■ **Fc fragment**

a product of papain hydrolysis of an immunoglobulin G consisting of parts of the constant regions of two heavy chains held together by a disulphide bridge, but excluding antigen-binding regions.

■ **feedback inhibition**

a form of metabolic control in which the end product of a pathway inhibits the enzyme, usually an allosteric enzyme, that catalyses the earliest irreversible reaction that is unique to the pathway. Also, the control a hormone exerts on the sequence of events that results in its synthesis and release, e.g. the action of thyroxine on secretion of thyrotropin (thryroid-stimulating hormone).

■ **feedback regulation**

control of a metabolic pathway by a metabolite of the pathway that acts in the direction opposite to metabolic flux, i.e. upstream or 'earlier' in the pathway.

■ **feed-forward regulation**

control of a metabolic pathway by a metabolite of the pathway that acts in the same direction as the metabolic flux, i.e. downstream or 'later' in the pathway, e.g. the activation of pyruvate kinase by fructose 1,6-bisphosphate.

■ **feline**

derived from cat.

■ **Ferguson plot**

a graphical representation of electrophoretic mobility as a function of gel concentration, according to the relationship: logMo/logM=KrC, where Mo is the free mobility, M is the mobility in a gel of concentration C, and Kr is a retardation constant related to molecular size.

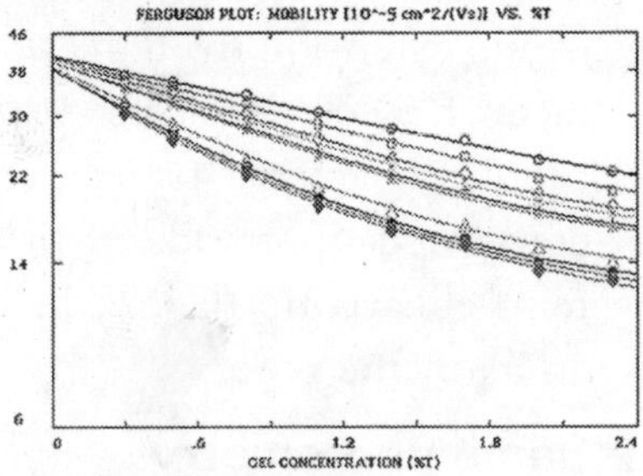

■ fermentation

the anaerobic degradation, usually by a micro-organism, of a sugar (or other source of energy) and biosynthetic intermediates, during which secondary metabolites may be produced.

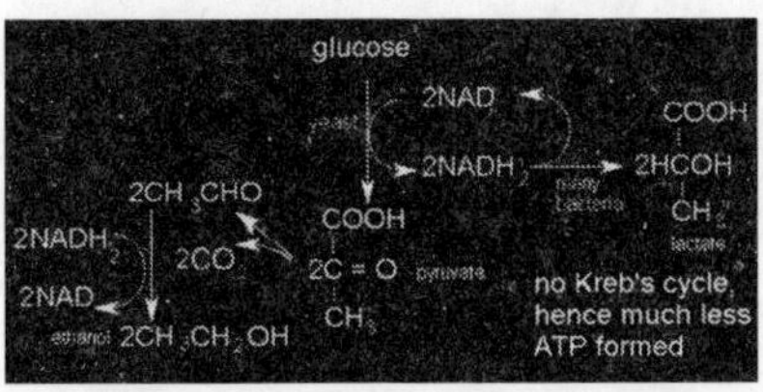

■ fibroblast

a relatively undifferentiated collagen-secreting cell of connective tissue.

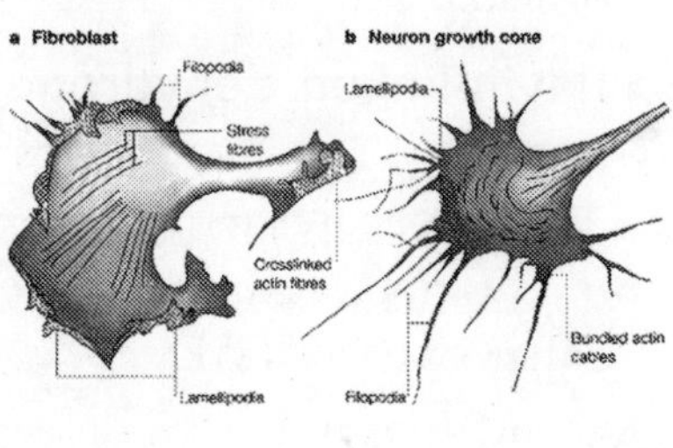

■ fibrous proteins

water-insoluble proteins that serve in a protective or structural role. The ratio of length to width generally exceeds 10.

■ fidelity

the property of an amino-acid-activating enzyme or a polymerase to correctly charge a tRNA or to correctly place a residue in a growing polypeptide or polynucleotide.

■ fingerprint

the unique pattern of ninhydrin-reactive spots found upon separation of a tryptic digest of a protein by paper electrophoresis in one dimension, followed by chromatography in a second dimension at a 908 angle. Also, any unique pattern that is generated by separation of fragments resulting from limited proteolysis of a protein or hydrolysis of a polynucleotide at specific cleavage sites.

■ finite decomposition

a type of mathematical modelling of steady-state metabolic rates according to the incremental contributions of all controlling factors.

■ first law of thermodynamics

energy can be neither created nor destroyed, only converted to other forms.

■ first-order kinetics

in enzymology, the direct dependence of rate on the substrate concentration; seen when this concentration is significantly below the Km.

■ Fischer convention

a representation of the absolute configuration of an asymmetrical carbon atom, especially the carbons of a sugar, in which the substituents further from the observer are shown above and below and the substituents closer to the observer are shown to the right and left.

■ Fischer projection formulae

formulas which, by convention, show bonds coming out of the plane drawn vertically, and bonds which project behind the plane are drawn horizontally.

■ flagellum

a long extension from the surface of a cell, often used for motility. Bacterial flagella are composed of strands of protein; eukaryotic flagella have an axoneme core.

■ flame photometry

an analytical technique to quantify small amounts of some elements, especially alkali metals, in solution; the solution is vapourised in a flame and the intensity of the characteristic spectral line emission produced by excitation of the metal atoms is detected by a photomultiplier.

■ flanking sequence

a DNA sequence located adjacent to a gene, either upstream from its 5'-end or downstream from its 3'-end.

■ flavin nucleotides

nucleotide coenzymes containing riboflavin, either FMN or FAD.

■ flavin-linked dehydrogenases

dehydrogenases requiring one of the riboflavin coenzymes, either FMN or FAD.

■ **flavoprotein**

an enzyme containing a FMN or FAD as a tightly bound prosthetic group.

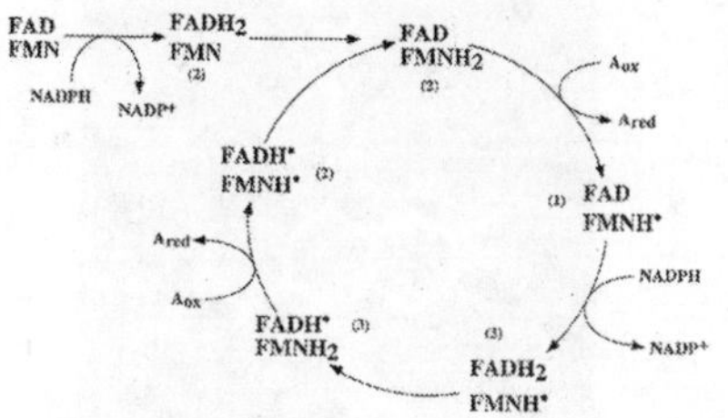

■ **flippase**

a putative enzyme that transports membrane elements, e.g. the lipopolysaccharide of Gram-negative bacteria, from the inner to the outer leaflet of the lipid bilayer cell membrane.

■ **florigen**

a hypothetical molecule that signals floral induction; i.e. the switch from the vegetative to the reproductive developmental pathway.

■ **flow cytometry**

the fluorescence-activated cell sorting technique applied to analysis and/or separation of cells by their physical and biological characteristics.

■ **fluid mosaic model**

a model proposed by Singer and Nicholson, describes biological membranes as a fluid lipid bilayer with embedded proteins creating a mosaic. Both protein and lipid components of the bilayer are able to move, but exhibit both structural and functional asymmetry.

■ **fluorescence**

the property of a compound or moiety of absorbing ultraviolet or visible light and re-emitting it nearly instantaneously at a longer wavelength; also the fluorescent emission itself.

Fluorescence Excitation and Emission

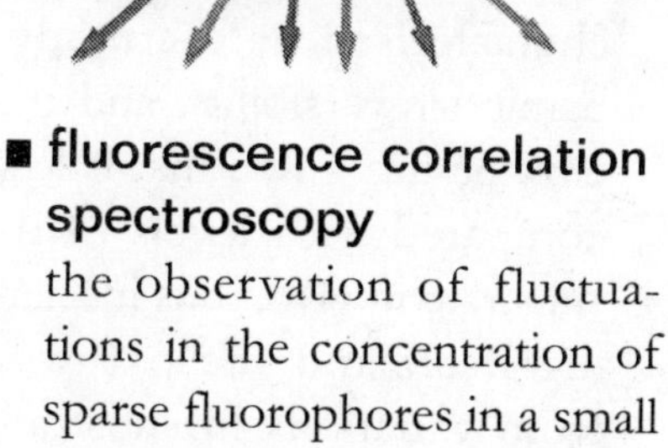

Figure 6

■ **fluorescence correlation spectroscopy**

the observation of fluctuations in the concentration of sparse fluorophores in a small observation area. Applications include estimation of transport coefficients as a fluorophore repopulates a bleached area, and of very low concentratons as individual molecules diffuse in and out of the experimental field of vision.

■ fluorography

a variant of autoradiography with improved sensitivity for detection of radioactivity of bands in slab gels. The weak emission from 3H, 14C or 35S excites a scintillant such as salicylate that in turn exposes the photographic film.

■ flux-generating step

an enzymic reaction that limits the rate of subsequent steps in its metabolic pathway. In a steady state the flux of metabolites, expressed as mmol/min, is the same for all steps. The step is usually characterised by thermodynamic irreversibility, and the enzyme has a Km that is low compared with levels of its substrate in vivo, thus insulating itself and the pathway from variations in substrate levels.

■ FMN (flavin mononucleotide)

riboflavin phosphate, a coenzyme of some redox enzymes.

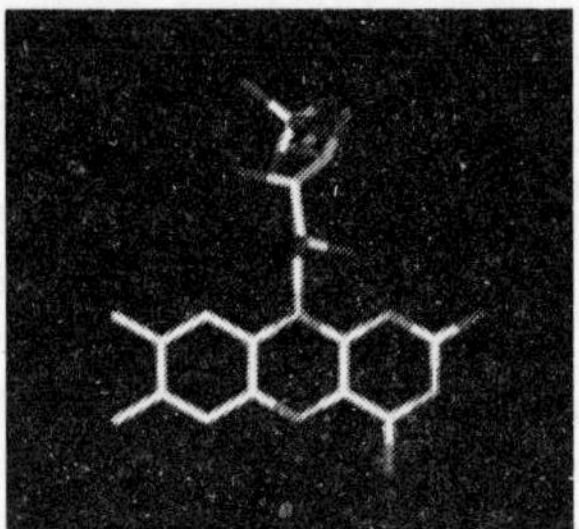

Flavin Mononucleotide (Vitamin B2)

■ folding unit

a nucleation site of protein naturation or renaturation.

■ Folin-Ciocalteu reagent

a reagent that uses phosphomolybdotungstate for the colorimetric detection of proteins.

■ footprinting

a group of techniques for identification of protein-binding sites on DNA. In dimethylsulphate (DMS) footprinting, for example, an end-labelled DNA fragment is partially methylated at guanidine residues with DMS, then exposed to the DNA-binding protein. The material is electrophoresed on a non-denaturing polyacrylamide gel to separate fragments that are methylated in the protein-binding sequence (and will no longer bind) from those that are methylated only elsewhere and will be less mobile due to the bound protein. The two gel bands are then eluted and cleaved with piperidine at the methylated sites. Comparison of the cleavage ladders of the two preparations will show unique bands due to methylation and cleavage at residues in the non-binding material that were essential for binding. In protection footprinting a DNA fragment with and without a binding protein is subjected to cleavage in non-binding areas, e.g. by partial methylation and cleavage by piperidine. The material is separated on a sequencing gel and the ladder of the protein-bound preparation will show a gap (the footprint) in its sequence where the protein has bound, compared with the ladder from 'naked' DNA. Protection footprinting may be applied to intact cells and compared with naked DNA in vitro as the control. In that case, ligand-mediated PCR is used to label and amplify desired areas of the DNA. Other methods of cleavage are by DNase I and by hydroxyl radicals (OH•).

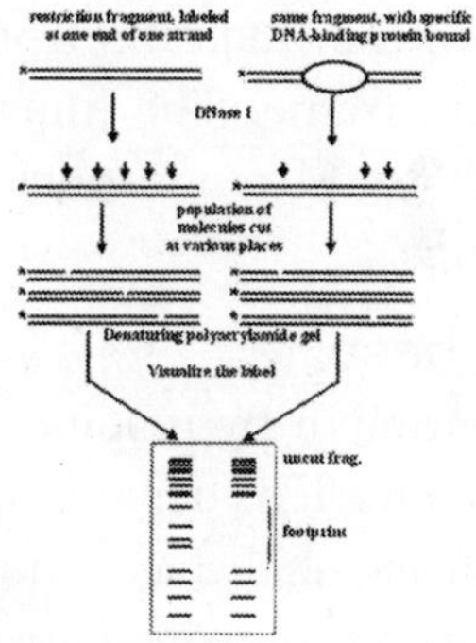

■ form

shaped like.

■ founder effect

the prominence in a population of a rare allele or haplotype due to the population's being historically isolated and having been descended from a small number individuals, among whom was one individual, the founder, who by chance possessed the rare allele or haplotype.

■ frame shift

a mutation caused by insertion or deletion of one or more paired nucleotides, changing the reading frame of codons and therefore the amino acids incorporated into proteins.

■ frameshift mutations

insertions or deletions of genetic material that lead to a shift in the translation of the reading frame. The mutation usually leads to nonfunctional proteins.

■ free base

the purine or pyrimidine moiety of a nucleoside, nucleotide or nucleic acid that is not attached to the pentose or pentose phosphate moiety.

■ free energy (G)

that portion of the total energy of a system that can do work at constant temperature and pressure.

■ free-energy change (ΔG')

the amount of free energy released (negative ?G) or absorbed (positive ?G) in a reaction at constant temperature and pressure.

■ freeze-fracture

a technique to prepare cells for scanning electron microscopy; cleavage of a solidly frozen cell, followed by freeze-etching, i.e. sublimation of some of the frozen water of the cell and coating the exposed subcellular structures with a film of platinum or carbon.

■ Fuelgen reaction

a qualitative colorimetric method for identification of DNA, especially in cytochemistry; treatment with fuchsin sulphurous acid to produce a red colour.

■ functional group

the specific atom or group of atoms that confers a particular chemical property on a biomolecule.

■ funnel concept

the new view of the protein folding problem in which the assortment of unfolded polypeptides are considered as an ensemble, the myriad of individual members of which achieve the native form by microscopically unique routes, all funneling down from the high-energy unfolded state to the low energy native conformation. If these molecules, on average, do not encounter an obstacle to refolding, the kinetics appears like the two-state case of the pathway model. Obstacles may be dead ends (kinetic traps) from which the miss-folded protein is extracted by Browninan forces or helped by a chaperonin to reverse to a higher energy state from which it can continue its folding. Another kind of bottleneck is a conformational entropy barrier, when there are only small differences in energy among the various possible conformations each folding polymer can achieve. It would then take some time, on the average, for the partially folded molecules to search the many possible conformations to find a continuation of the folding sequence. By viewing the energy co-ordinate as an energy landscape, folding of specific unfolded proteins may be viewed as passing over plateaus, down steep ravines or trapped in moats. Although the funnel concept, or 'new view' of protein folding was proposed as an alternative to the hypothesis that each protein has only one folding pathway (the 'old view), there in fact may be multiple pathways, but they share many features.

■ furanose

the form of a sugar when it is condensed into a five-membered ring. It consists of four carbon atoms and the oxygen atom that is the link to the anomeric carbon atom.

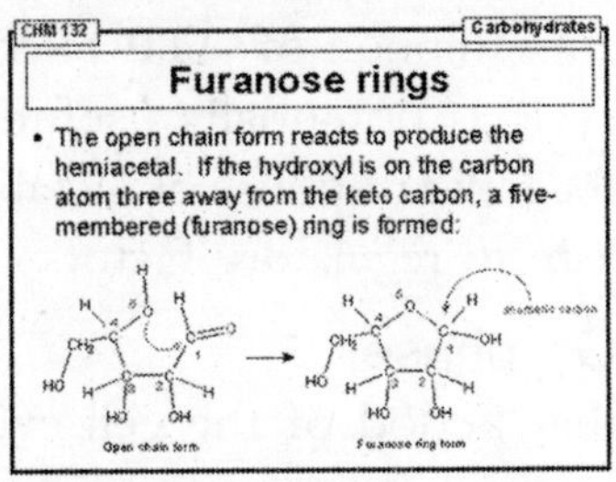

■ fusion protein

depending on the context means one of two different things. Most commonly, refers to the protein product of a gene created by the fusion of two distinct genes, or portions of genes. Less commonly refers to one of a family of proteins that participate in membrane fusion.

■ **futile cycle (substrate cycle)**
the coupling through common substrate/product pairs of an energy-requiring enzymic reaction with another energy-producing reaction that regenerates one of the products of the first reaction. The net result is the expenditure of energy, e.g. the phosphofructokinase and fructose phosphate phosphatase reactions that together generate and hydrolyse fructose bisphosphate with the net consumption of ATP. Such a cycle is potentially useful in the rapid response of metabolism to regulatory factors.

■ **G1 phase**
that period of the cell cycle in which preparations are being made for chromosome duplication, which takes place in the S phase.

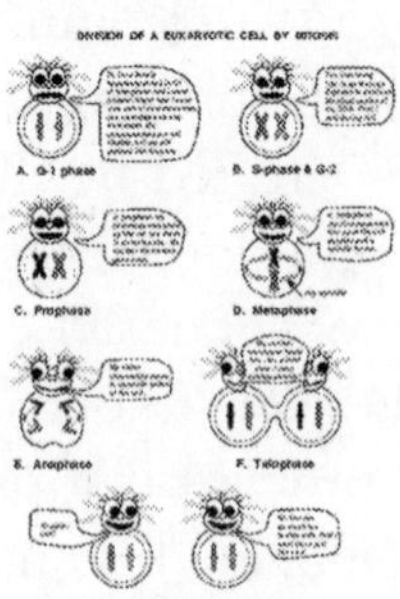

■ **G2 phase**
that period of the cell cycle between S phase and mitosis (M phase).

■ **gametes**
the ova and the sperm, haploid cells that unite during fertilisation to generate a diploid zygote.

■ **gangliosides**
sphingolipids containing complex oligosaccharides as head groups.

■ **gap junction (cell-to-cell channel)**
a connection between adjacent cells that allows small molecules and ions, e.g. amino acids, sugars and nucleotides, to pass from the cytoplasm of one cell to the cytoplasm of the other.

■ **gastrulation**
during early embryogenesis, the invagination and reshaping of the cells of the blastoderm that results in differentiation into ectoderm, endoderm and mesoderm.

■ **gate**
a regulated channel through which ions pass across the

plasma membrane; gating is the opening and closing of the channel.

■ **gatekeeper protein**

a protein that monitors transfer of a protein from the endoplasmic reticulum to the Golgi apparatus and prevents transfer of newly synthesised proteins with inappropriate conformations or with unpaired thiol groups.

■ **gauche conformation**

an arrangement in space of the carbon-carbon bonds of the alkyl chains of a fatty acid residue of a membrane lipid. The backbone of the chain is rotated 120° from the trans conformation, to force the backbone into making a 60° turn through the bond.

■ **GDNF**

glial cell line-derived neurotrophic factor.

■ **gel filtration**

a chromatographic procedure for the separation of a mixture of molecules on the basis of size.

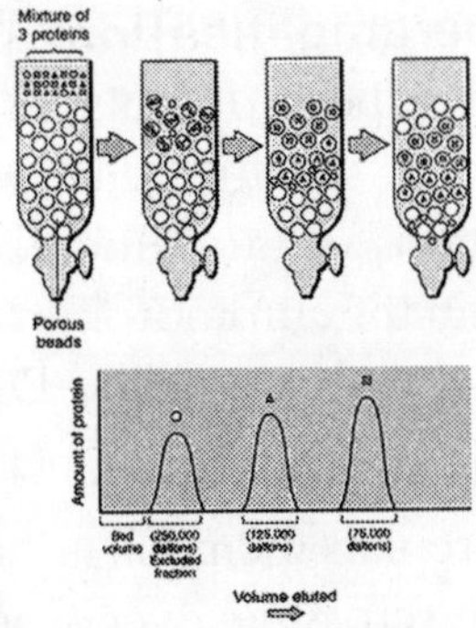

■ **Gel fitration chromatography**

a technique that makes use of certain polymers that can form porous beads with varying pore sizes.In columns made from such beads, it is possible to separate molecules, which cannot penetrate beads of a given pore size, from small molecules that can.Also called gel-exclusion or molecular seive chromatography.

■ **gene**

in genetics, a unit inferred from the pattern of inheritance; in molecular biology, defined narrowly as a section of DNA that is expressed as RNA or, more widely, as a coding sequence of DNA and associated regulatory sequences.

■ **gene amplification**

the synthesis of multiple copies of a specific gene, which remain as tandem repeats within the chromosome or are segregated as satellite DNA.

■ **gene conversion**

the replacement of an allele during crossing- over owing to mismatch repair of the heteroduplex intermediate; rather than being equally represented in daughter DNA strands, one allele is over-represented and the other is under-represented. Partial repair can result in a hybrid or mosaic gene.

■ **gene duplication**

the inherited appearance of multiple copies of a gene. Duplication may result in identical or defective copies, either as tandem repeats at the original chromosomal locus, at a neighbouring locus or at a completely different locus. Duplication may be benign or may be the basis of a pathology. If benign, one of the copies will be freed of any selective pressure and may evolve a new function.

■ **gene expression**

the production of a biologically active gene product, either constitutive or regulated.

■ **gene product**

the result of gene expression, i.e. an RNA or protein molecule.

■ **gene splicing**

the enzymatic attachment of one exon to another.

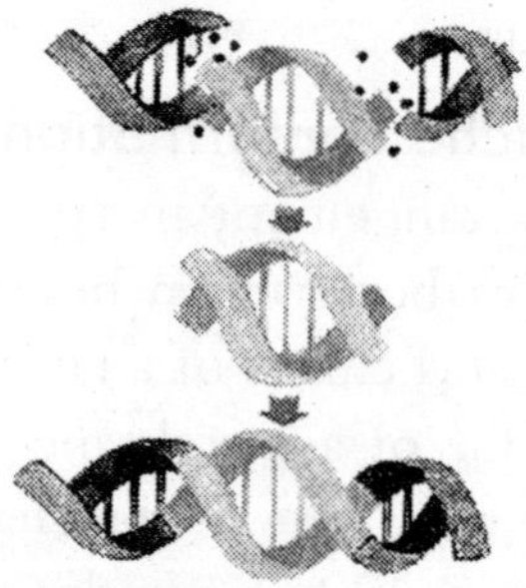

■ **gene targeting**

the introduction of a homologous DNA sequence into a specific site in the genome of a cell, either by replacement of the former sequence, i.e. sequence replacement, or by insertion into the former sequence, i.e. sequence insertion. A vector that is homologous and co-linear with a partial sequence of the targeted gene is introduced into an appropriate cell, e.g. a stem cell of some sort; it is incorporated into the genome of some of the cell's progeny and replaces the former sequence. If, however, homologous regions of the vector hybridise to the gene, and it is then incorporated into the gene by homologous recombination, the result is a disruption of the gene by insertion of the vector sequence.

■ gene therapy
the in vivo introduction of a functional gene, and its expression, in the germ line cells (germ line therapy) or somatic cells (somatic gene therapy) of an individual who does not possess the normal gene.

■ general acid
in chemistry, a proton donor that can participate in catalysis.

■ general acid-base catalysis
catalysis involving a proton transfer from/to a molecule other than water. Compare to specific acid-base catalysis.

■ general recombination
recombination that occurs between homologous chromosomes at homologous sites.

■ generation time
the time it takes for a cell to double its mass under specified conditions.

■ genetic code
the set of triplet code words in mRNA or DNA, coding for the amino acids of proteins.

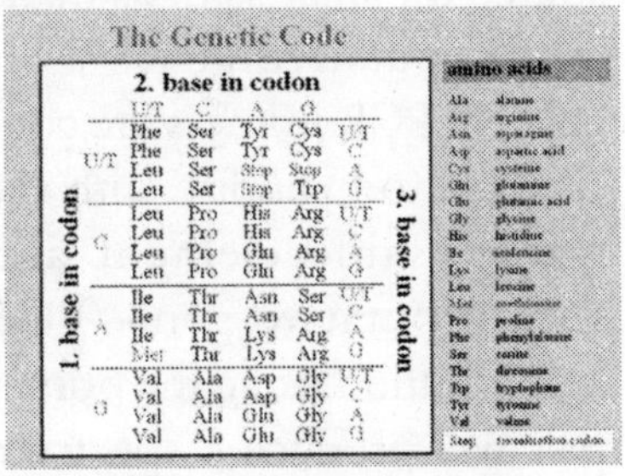

The Genetic Code

1. base in codon	2. base in codon: U/T	C	A	G	3. base in codon
U/T	Phe	Ser	Tyr	Cys	U/T
	Phe	Ser	Tyr	Cys	C
	Leu	Ser	Stop	Stop	A
	Leu	Ser	Stop	Trp	G
C	Leu	Pro	His	Arg	U/T
	Leu	Pro	His	Arg	C
	Leu	Pro	Gln	Arg	A
	Leu	Pro	Gln	Arg	G
A	Ile	Thr	Asn	Ser	U/T
	Ile	Thr	Asn	Ser	C
	Ile	Thr	Lys	Arg	A
	Met	Thr	Lys	Arg	G
G	Val	Ala	Asp	Gly	U/T
	Val	Ala	Asp	Gly	C
	Val	Ala	Glu	Gly	A
	Val	Ala	Glu	Gly	G

amino acids	
Ala	alanine
Arg	arginine
Asn	asparagine
Asp	aspartic acid
Cys	cysteine
Gln	glutamine
Glu	glutamic acid
Gly	glycine
His	histidine
Ile	isoleucine
Lys	lysine
Leu	leucine
Met	methionine
Pro	proline
Phe	phenylalanine
Ser	serine
Thr	threonine
Trp	tryptophane
Tyr	tyrosine
Val	valine
Stop:	termination codon

■ genetic fingerprinting
a technique for establishing genetic relationships, especially in forensic medicine, by comparison of the occurrence of uncommon genetic markers.

■ genetic footprinting
an approach to the identification of the function of a large number of putative genes of a micro-organism, such as may have been uncovered by genome sequencing. A transposable element is inserted

into a large number of sites in the genome of the micro-organism, and the mutagenised population is then grown under a wide variety of conditions that may suggest a gene's function. After many population doublings under each set of conditions, the micro-organisms' DNA is extracted and used as a resource for analysis of as many of the putative genes as is desired. PCR primers are constructed to hybridise with the transposable element and with a putative gene. PCR amplification using the primer pair will produce a spectrum of bands on separation by polyacrylamide-gel electrophoresis, one from each mutation that can grow under the set of conditions, i.e. a genetic footprint of the gene. A comparison of the gene's footprints under permissive and restrictive conditions will indicate, by the absence of bands in the latter footprint, those growth conditions that require the function of the gene.

■ genetic information

the hereditary information contained in a sequence of nucleotide bases.

■ genetic map

a diagram showing the position of specific genes along a chromosome relative to markers.

■ genome

the genetic complement of an organism represented by its DNA or, in some viruses, its RNA.

■ genomic DNA

DNA that has been isolated from a cell and therefore contains introns, as opposed to cDNA.

■ genotype

the genetic information of an organism, as distinct from its physical expression (phenotype).

■ germ-line cell

a class of animal cells formed early in embryogenesis that are set aside for reproductive function (give rise to gametes). All other cells are somatic cells.

■ **globular**

descriptive of the folding of a protein upon itself in several convolutions to create interactions of the side chains in many weak salt bridges, hydrogen bonds and hydrophobic bonds that result in a roughly spherical (globular) molecular shape.

■ **globular proteins**

soluble proteins with a globular (somewhat rounded) shape. Their axial to diameter ratio is less that 10:1, and usually closer to 1:1.

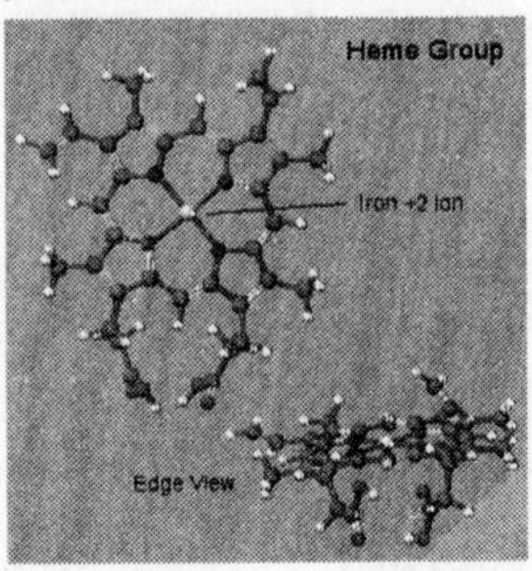

■ **globulin**

archaic nomenclature for a protein that is sparingly soluble in water but is soluble in dilute salt solutions. Euglobulins do not dissolve in salt-free water, whereas pseudoglobulins are soluble in salt-free water. Gamma (γ)-globulins are a population of globulins defined further by their electrophoretic behaviour, and include the immunoglobulins.

■ **glucocorticoid**

an adrenal steroid that increases blood glucose concentration, e.g. cortisol.

■ **glucogenic amino acids**

amino acids with carbon chains that can be metabolically converted into glucose *via* gluconeogenesis. Contrast with ketogenic amino acids.

■ **gluconeogenesis**

the production of sugars from nonsugar precursors such as lactate or amino acids. Applies more specifically to the production of free glucose by vertebrate livers.

■ **glucose effect**

the repression of some bacterial enzymes by the presence of glucose in the growth medium. Depression of cyclic AMP levels due to the availability of glucose is mediated by inhibition of adenylate cyclase by a glucose catabolite. In the absence of glucose, a complex between cyclic AMP and its binding protein attaches to specific regions of DNA and activates transcription. Cyclic AMP acts as a starvation signal, as it does in animal cells, but in bacteria it acts at the transcriptional level.

■ **glycan**

a general term for a polymer of monosaccharide units joined by glycosidic bonds, may or may not have other components, e.g. peptidoglycan.

■ **glycation**

the reversible condensation of the carbonyl group of a sugar with an amino group of a protein to form a Schiff base that often triggers subsequent irreversible rearrangements, e.g. the condensation of glucose with haemoglobin A to form haemoglobin A_{1c}.

■ **glycerophospholipid**

an amphipathic lipid with a glycerol backbone. A (usually saturated) fatty acid is esterified to C-1 and a (usually unsaturated) fatty acid is esterified to C-2 of glycerol. A polar alcohol is attached through a phosphodiester linkage to C-3.

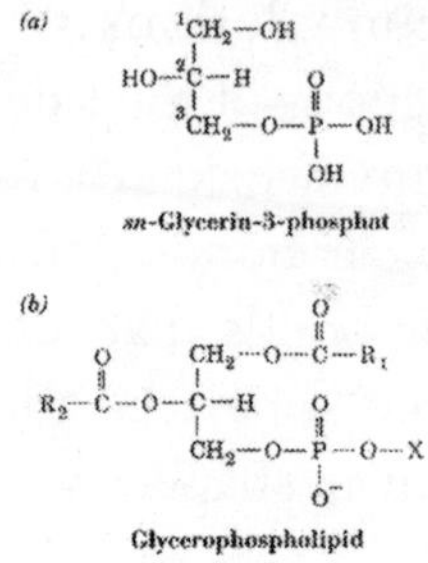

■ **glycobiology**

the study, at the molecular level, of the structural and functional roles of carbohydrate-containing structures and of the interactions of carbohydrates with other structures, especially proteins.

■ **glycocalyx**

the carbohydrate coating on the external surface of a cell membrane.

■ **glycoform**

one of several variants of a glycoprotein that have identical polypeptide moieties but differ in the site of glycosylation (site heterogeneity) and/or the nature of the oligosaccharide moiety. Glycoforms are characteristic of the tissue of origin, or of the disease or mutagenic state.

■ **glycogen**

a polymer of glucose residues in 1,4 linkage, with 1,6 linkages at branchpoints.

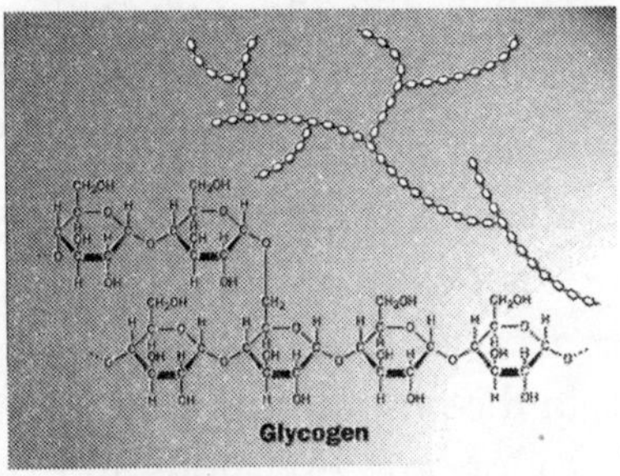

■ **glycogenic**

describing amino acids whose metabolism may lead to gluconeogenesis.

■ **glycolipid**

a compound with both lipid and sugar moieties, e.g. lipopolysaccharide of Gram-negative bacteria cell walls, blood group antigens, gangliosides (carbohydrate moieties attached to N-fatty acylsphingosine), cerebrosides and ceramides.

■ **glycolysis**

the catabolic pathway by which a molecule of glucose is broken down into two molecules of pyruvate, which undergoes further metabolism in other pathways.

■ **glycoprotein**

a protein containing (a) carbohydrate sidechain(s) in either an N- or O-glycosidic linkage.

■ **glycoprotein hormone**

a representative of a class of hormones that share nearly identical A-chains and unique B-chains, each with about 125 amino acid residues and N-linked or O-linked carbohydrate moieties, e.g. thyrotropin, follitropin.

■ **glycosaminoglycan**

a molecule composed of alternating monsaccharide units. One is either GlcNAc or GalNac, the other is a usually glucuronic acid. Formerly called mucopolysaccharide.

■ **glycosidic bonds**

bonds between a sugar and another molecule (typically an alcohol, purine, pyrimidine, or a second sugar). Divided into N-glycosidic or O-glycosidic linkages, depending on the point of attachment.

glyoxylate cycle

a variant of the citric acid cycle present in bacteria and some plant cells which is used for the net conversion of lipid into carbohydrate.

glyoxysome

a specialised peroxisome containing the enzymes of the glyoxylate cycle. Particularly prevalent and important in germinating seeds.

Goldman equation

an equation expressing the quantitative relationship between the concentrations of charged species on either side of a membrane and the resting transmembrane potential.

Golgi apparatus

a stack of flattened vesicles that functions in the posttranslational processing and sorting of proteins. The Golgi receives proteins from the rough endoplasmic reticulum (RER) and directs them to secretory vesicles, lysosomes or the cell's plasma membrane. The movement of the proteins takes place by transfer vesicles that bud off from the RER or Golgi and fuse with the Golgi, lysosomes or plasma membrane. The surface of the Golgi that faces the RER is termed the cis-Golgi, the surface towards the apical membrane is the trans-Golgi, and the intermediary part is the medial Golgi.

Golgi complex

a complex and dynamic membranous organelle of eukaryotic cells. Golgi bodies function in the posttranslational modification (glycosylation) and secretion of proteins or insertion into membranes. Composed of cis, medial and trans compartments as well as a less well defined "trans-golgi network."

Gortner and Grendel model

a model in which erythrocyte membranes comprise only phosphoacylglycerols, i.e. no protein or sterol component is envisaged; the model presents a bilayer in which each leaflet is composed of lipids stacked perpendicular to the plane of the membrane, with their alkyl chains oriented inwards (towards the opposing leaflet) and with the polar groups oriented towards the exposed surfaces.

gram molecular weight
for a given compound, the weight in grams that is numerically equal to its molecular weight.

gram-negative
descriptive of bacteria that do not stain by the Gram method, i.e. bacteria with two membranes.

gram-positive
descriptive of bacteria that stain by the Gram method, i.e. bacteria with single membranes.

grana
stacks of thylakoids in chloroplasts.

granulocyte
a granular leucocyte; a basophil, eosinophil or polymorphonuclear leucocyte (neutrophil).

granzyme
a T-cell (granulocyte) proteinase.

Greek key
a topology found in some protein structures; antiparallel - strands are organised into a barrel in which strands that are adjacent in the primary structure are not always adjacent in super-secondary structure but, when represented in two dimensions, form a Greek key pattern.

green beard
a conspicuous phenotype that allows behavioural differentiation among individuals of the species. When the consequent behaviour is beneficial, the encoding gene can be considered a selfish gene.

ground state
the normal, stable form of an atom or molecule. Contrast with the excited state.

group transfer potential
a measure of the ability of a compound to donate an activated group (such as a phosphate group). Computed and expressed from the standard free energy of hydrolysis.

growth cone
in a developing or regenerating neuron, the leading edge of a focus of dendrite extension, which is responsive to environmental cues that deter-

mine its growth rate and the direction of that growth.

■ **growth factor**

a protein that binds to receptors on specific cells and promotes their growth, e.g. nerve growth factor, epidermal growth factor, platelet-derived growth factor.

■ **growth fork**

the region on a DNA duplex molecule where synthesis is taking place.It resembles a fork in shape, since it consists of a region of duplex DNA connected to a region of unwound single strands.

■ **guanine**

a purine base found in DNA or RNA.

■ **guanosine**

a purine nucleoside found in DNA and RNA.

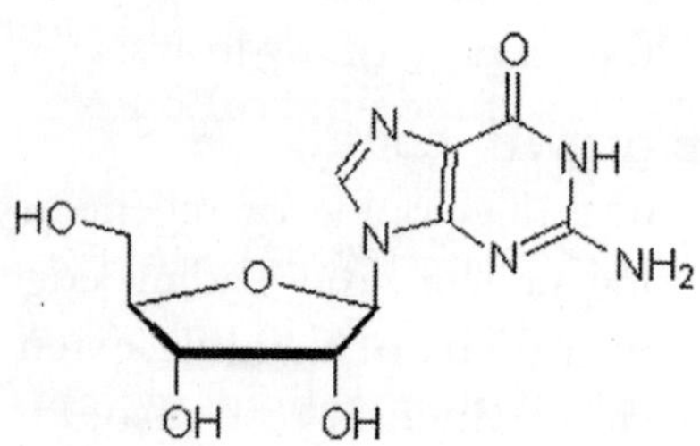

■ **guide sequence**

in the self-splicing of RNA, the purine-rich region of the intron that binds a pyrimidine-rich region of the upstream exon to the site where transesterification takes place.

■ **haemolysis**

the bursting of erythrocytes due, for example, to a hypotonic environment.

■ **hairpin loop**

a single-stranded complementary region of DNA or RNA that folds back on itself and base-pairs into a double helix.

■ **half-life**

the time required for the disappearance or decay of one-half of a given quantity of substance.

■ **halophilic**

descriptive of organisms that survive and grow only in high salt concentrations, e.g. >3 M NaCl.

■ **handshake**

an interaction of two identical or nearly identical structures, proteins or nucleic acids, which fit each other in a symmetrical fashion.

■ **haplo-**

single.

■ **haploid**

having a single set of genetic information; describing a cell with one chromosome of each type; used in reproduction to prevent polyploidy.

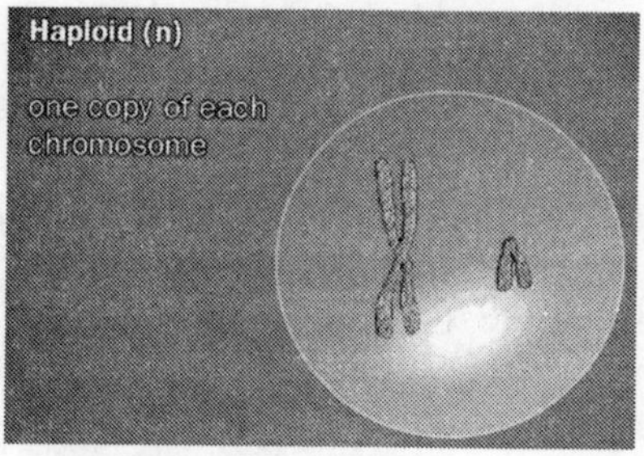

■ **haploid cell**

a cell containing only one chromosome of each type.

■ **haplotype**

a characteristic combination of alleles on a single chromosome, which may persist in a population because of their close proximity on the chromosome, a founder effect or natural selection.

■ **hapten**

a small molecule which by itself is not an antigen, but which as a moiety of a larger structure (a haptenic determinant) can serve as an antigenic determinant.

■ **hapto-**

bind to.

■ **haptoglobin**

serum protein which binds free haemoglobin for transport to the liver and destruction.

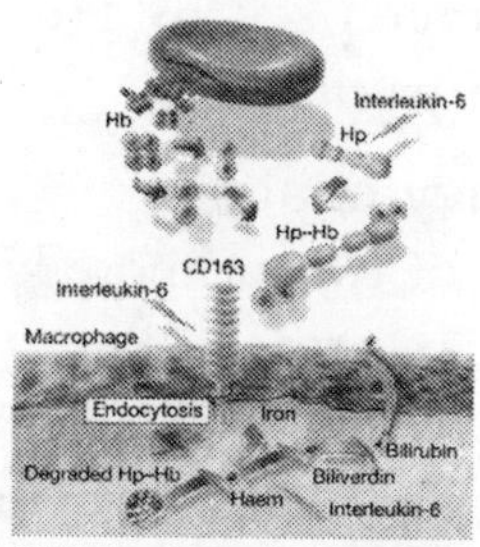

The Hp–Hb complex is engulfed by the macrophage and digested to release haem

■ **Harden-Young ester**

obsolete name for fructose 1,6-bisphosphate.

■ **Haworth perspective formulas**

a method for representing cyclic chemical structures (sugars) in order to define the configuration around each chiral center.

■ **Haworth projection**

a two-dimensional representation of pyranose and furanose structures in which they are shown as hexagons and pentagons respectively. The lower ring atoms are understood as being towards and the upper ones away from the observer; the substituents of ring carbon atoms are shown directly above and below the apices of the polygon.

■ H-DNA
a form of triplex DNA characterised by Hoogstein base pairing.

■ heavy chain
one of two or more kinds of polypeptide chain of a heteromultimeric protein, distinguished by molecular mass, e.g. one of the larger polypeptide chains that, when combined with another heavy chain and two light chains, make up an immunoglobulin molecule; also the largest of the polypeptide chains of any multimeric protein with non-identical subunits.

■ heavy isotopes
forms of atoms that contain greater numbers of neutrons than the most common form (e.g., ^{15}N, ^{13}C).

■ Heinz body
a precipitate in the cytoplasm of an erythrocyte of a spontaneously oxidised, unstable, haemoglobin.

■ helicase
an enzyme that catalyses the unwinding of strands in a DNA molecule before replication.

■ helix
a spiral structure with a repeating pattern.

■ helix cap
a sequence of amino acids in a protein that terminates a stretch of α-helical structure.

■ haeme
an iron-porphyrin complex found in haemoglobin and cytochromes.

■ haeme protein
a protein containing a haeme as a prosthetic group.

■ hemi-
half.

■ hemiacetal
the product of reversible condensation of an aldehyde and an alcohol in which the alcoholic hydroxy group adds

across the carbonyl group of the aldehyde, e.g. glucose acting as both an aldehyde and an alcohol to give the internally condensed glucopyranose; also the bond formed by this condensation. An acetal is the product formed by abstraction of the hydroxy group of a hemiacetal and the hydrogen of a second alcohol, e.g. a glucopyranoside or a polysaccharide.

■ hemiketal

the product of reversible condensation of a ketone and an alcohol in which the alcoholic hydroxy group adds across the carbonyl group of the ketone, e.g. fructose acting as both a ketone and an alcohol to give the internally condensed fructofuranose; also the bond formed by this condensation. A ketal is the product formed by abstraction of the hydroxy group of a hemiketal and the hydrogen of a second alcohol, e.g. a fructofuranoside.

■ hemisubstrate

an enzyme inhibitor that acts in the initial phases of enzyme action as a substrate, but does not complete the catalytic cycle to regenerate an active enzyme; e.g. organophosphate ester anticholinesterases, which mimic acetylcholine in that the phosphoric acid moiety of the hemisubstrate acts like the acetly group of acetylcholine to form an ester with the enzyme's active-site serine residue but, unlike the acetyl-enzyme intermediate of the normal enzymic action, the phosphoryl-enzyme is only very slowly hydrolysed to regenerate an active enzyme.

■ haemoglobin

a haeme protein of erythrocytes which functions in oxygen transport.

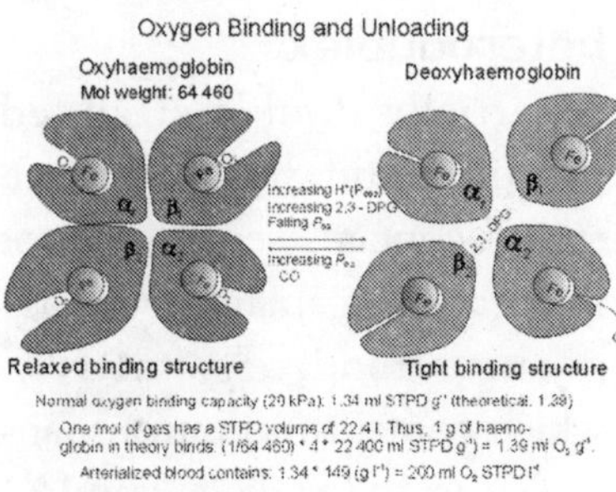

■ Henderson-Hasselbalch equation

a logarithmic form of the equation for the dissociation of a weak acid: pH=

pKa+log[salt form]/[acid form].

hepat
liver.

hepatocyte
the major cell type of liver tissue, which is parenchymal tissue. Liver also contains fixed macrophages called Kupfer cells and other cells.

hept-
seven.

hetero-
different.

heterochromatin
highly condensed regions of chromosomes that are not usually transcriptionally active.

heteroduplex
a partially double-stranded polynucleotide in which one strand contains sequences not fully complementary to the opposite strand, e.g. mRNA, which contains no introns, annealed to the coding strand of its corresponding gene, which contains introns. Heteroduplexes are products of recombination between DNA duplexes at a region of heterology.

heteroduplex DNA
duplex DNA containing complementary strands derived from two different DNA molecules with similar or partially identical sequences. Sometimes from two unmanipulated species, other times a gene has been modified and reintroduced.

heterogenous
descriptive of a phenotype which is expressed by two or more genotypes, in contrast with polygeneous, which is descriptive of a phenotype which requires several genes together for its expression.

heterolytic cleavage
the splitting of a covalent bond that leaves both bonding electrons with one of the atoms.

heterophagy
the action of a lysosome of a phagocytic cell to digest extracellular materials.

heteropolymer
a polymer containing more than one type of monomeric unit.

■ **heteropolysaccharide**
a polysaccharide containing two or more types of sugars.

■ **heterotroph**
an organism that requires complex nutrient molecules, such as glucose, as an energy source.

■ **heterotropic enzyme**
an allosteric enzyme requiring a molecule other than its substrate as a modulator.

■ **heterozygous**
describing an organism (a heterozygote) that carries two different alleles for a given gene.

■ **hex-**
six.

■ **hexose**
a simple sugar with a backbone containing six carbon atoms.

■ **high-energy bond**
a chemical bond whose hydrolysis results in the generation of greater than 30kJ (7kcal) of energy or, if coupled to an energetically unfavourable reaction, can drive that reaction forward.

■ **high-energy compound**
a compound that undergoes hydrolysis with a high negative standard free energy change.

■ **high-performance liquid chromatography (HPLC)**
a type of chromatography using relatively high pressures and small diameter column packings to achieve sharp and highly reproducible elution profiles. Used to be called high pressure liquid chromatography.

■ **hinge**
a flexible polypeptide sequence connecting two domains of a protein that can move with respect to each other.

■ **histones**
the family of five basic proteins (H1, H2A, H2B, H3 and H4) that associate tightly with DNA in the chromosomes of all eukaryotic cells.

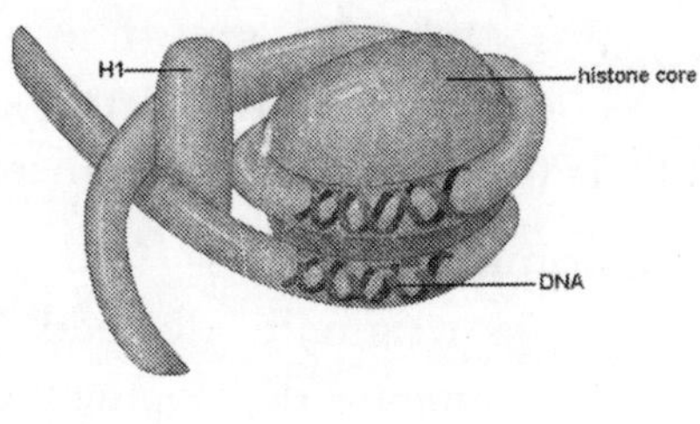

■ holliday intermediate

an intermediate in genetic recombination in which two double-stranded DNA molecules are joined by virtue of a reciprocal crossover involving one strand of each molecule.

■ holoenzyme

a catalytically active enzyme including all necessary subunits, prosthetic groups, and cofactors.

■ homeobox

a 180 base pair, highly conserved DNA sequence encoding a protein domain (the homeodomain) found in many proteins that play a regulatory role in transcription. Particularly important in development.

■ homeostasis

the maintenance of a dynamic steady state by regulatory mechanisms that compensate for changes in circumstances. For example, maintaining body temperature whether in a 20°C or a 4°C environment.

■ homeotic genes

genes originally defined as regulating the development of the pattern of segments in the Drosophila body plan. Similar genes are now known to occur in most vertebrates.

■ homo-

same.

■ homologous chromosomes

chromosomes that carry the same pattern of genes, but not necessarily the same alleles.

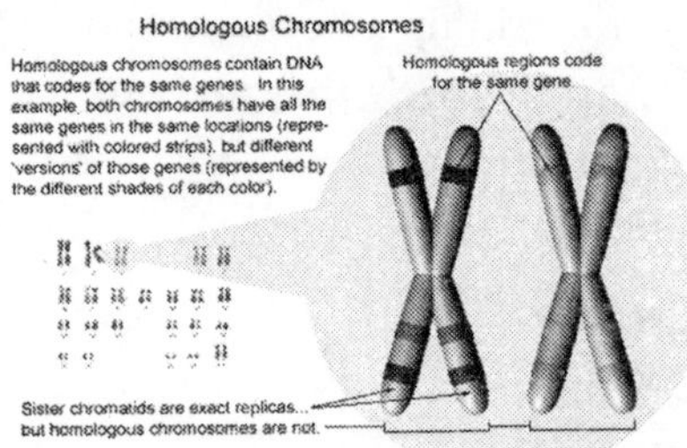

■ homologous genetic recombination

recombination between two DNA molecules of similar sequence, occurring in all cells; occurs during meiosis and mitosis in eukaryotes.

■ homologous proteins

proteins from different species having similar sequences and functions similar in each. For example the many species which have haemoglobin which transports oxygen in all.

■ **homopolymer**

a polymer composed of only one type of monomeric building block.

■ **homopolysaccharide**

a polysaccharide made up of only one type of monosaccharide unit.

■ **homotropic enzyme**

an allosteric enzyme that is modulated by its substrate.

■ **homozygous**

describing an organism (a homozygote) that carries two identical alleles for a given gene.

■ **hormone**

a chemical substance (peptide, steroid) synthesised in small amounts by an endocrine tissue and carried in the blood to another tissue, where it acts as a messenger to regulate the function of the target tissue.

■ **hormone receptor**

a protein in, or on the surface of the target cells of a tissue that functions as a sensor for the hormone by binding the hormone and initiating a cellular response.

■ **hormone receptor**

a protein that is located on the cell membrane or inside the responsive cell and that interacts specifically with the hormone.

■ **host cell**

a cell used for growth and reproduction of a virus.

■ **Hunter and Downs plot**

a graphical method for determination of dissociation constants of enzyme inhibitors. Where I is total inhibitor concentration and α is the ratio of inhibited to non-inhibited rates, a plot of $I\alpha/(1-\alpha)$ on the ordinate (y-axis) against substrate concentration on the abscissa (x-axis) will give a line that intersects the ordinate at Ki; if inhibition is competitive, the slope of the plot is Ki/Km, if non-competitive, the slope is zero. The advantage of this method lies in its not requiring that data be collected at constant substrate or inhibitor concentration.

■ **hvbrid (or chimeric)**

plasmid

a plasmid that contains DNA from two different organisms.

■ hydrogen bond

a weak electrostatic attraction between one electronegative atom (biochemically oxygen or nitrogen) and a hydrogen atom covalently linked to a second oxygen or nitrogen atom.

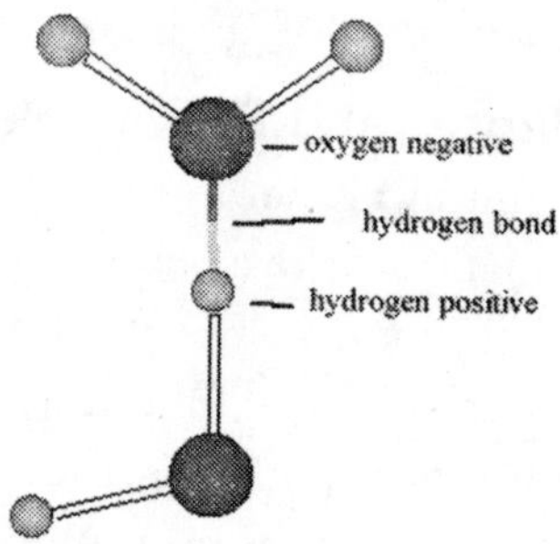

■ hydrolases

enzymes that catalyse hydrolysis reactions. Includes proteases, lipases, phosphatases, and nucleases.

■ hydrolysis

cleavage of a bond, such as a peptide bond, by the addition of the elements of water, yielding two or more products.

■ hydronium ion

the ion H_3O^+.

■ hydropathy index

a scale that expresses the relative hydrophobic and hydrophilic tendencies of amino acid R groups. Used to predict membrane-spanning regions.

■ hydrophilic

polar or charged molecules that dissolve readily in water.

■ hydrophobic

preferring not to be in contact with water, as is the case with the hydrocarbon portion of a fatty acid or phospholipid chain.

■ hydrophobic bond

the association of non-polar side chains of a protein that is driven by minimisation of the relatively unfavourable interactions of water molecules with the non-polar groups, and maximisation of the favourable interaction of water molecules with themselves when they are removed from contact with the non-polar residues and are added to the bulk water phase.

■ hydrophobic effect

the noncovalent association of nonpolar groups with each other in aqueous solution.

■ **hydrophobic interactions**
the association of nonpolar groups or compounds with each other in aqueous systems due to their insolubility in water.

■ **hydrophobic zipping**
a view of the initiation of protein folding in which the formation of hydrophobic interactions between very close groups bring slightly more remote groups into interaction distance in an accelerating condensation.

■ **hydroxyapatite**
a calcium phosphate gel used, in the case of nucleic acids, to selectively absorb duplex DNA-RNA from a mixture of single-stranded and duplex nucleic acids.

■ **hyper-**
above, excessive.

■ **hyperchromic effect**
the large increase in light absorption at 260 nm occurring as a double-helical DNA is melted.

■ **hypervariable region**
a localised sequence of a variable region of an immunoglobulin that shows a particularly large variation in amino acid sequence compared with all other immunoglobulins; presumably the part of the immunoglobulin that interacts directly with the antigen.

■ **hypo-**
below, deficient.

■ **ia**
got.

■ **iasis**
full of.

■ **idiotype**
a classification of immunoglobulin molecules according to the antigenicity of the variable regions. Each idiotype is unique to a particular immunoglobulin raised to a particular antigen.

■ **ile**
little version.

■ **illa**
little version.

■ **illegitimate recombination**
recombination with non-homologous regions of DNA. Each of the polynucleotide strands of double-stranded

DNA is religated with DNA from different regions of the same or different chromosomes.

illus

little version.

im-, in-

not.

image plate

a kind of area detector; a phosphor-impregnated plate that can be scanned to quantify its radiation exposure, and can then be regenerated.

imino group

a secondary amine, e.g. the ;NH group of proline.

immune response

the generation of antibodies to an antigen by a vertebrate organism.

immunofluorescence

a cytological technique in which a specific fluorescent antibody is used to label an antigen. Frequently used to determine the location of an antigen in a tissue or a cell.

immunoglobulin

a protein made in a B plasma cell and usually secreted; it interacts specifically with a foreign agent. Synonymous with antibody. It is composed of two heavy and two light chains linked by disulphide bonds. Immunoglobulins can be divided into five classes (IgG, IgM, IgA, IgD, and IgE) based on their heavy-chain component.

imposter

a gene product that can replace a missing homologous protein. Knockout experiments are sometimes misleading because a physiological function is preserved by an imposter; in such cases, a lethal mutation can produce a more informative phenoytpe than the knockout.

imunodiffusion

a method for testing the identity of several antigens using an Ouchterlony plate, a glass plate covered with a thin layer of agar on which is incised a hole surrounded by several other holes that serve as wells. In the central well is placed an antibody, and the antigens in question are placed in the surrounding wells; the antibody and antigens diffuse towards each other until they meet and form a visible precipitin line. An antigen in an adjacent well forms a precipitin line that is continuous with the first only if the two antigens are identical. Similarly, the plate can be used to test the identity of antibodies against a single antigen in the central well.

■ **in**

stuff.

■ **In vitro**

literally, "in glass," describing whatever happens in a test tube or other receptacle, as opposed to what happens in whole cells of the whole organism (*in vivo*).

■ **in vivo**

literally, "in life" meaning in a living cell or organism.

■ **inborn error**

an inherited defect in a metabolic enzyme or other gene product.

■ **inclusion body**

an amorphous deposit in the cytoplasm of a cell; an aggregated protein appropriate to the cell but damaged, improperly folded or liganded, or a similarly inappropriately processed foreign protein, such as a viral coat protein or recombinant DNA product.

■ **indirect end labelling**

a method to locate, between restriction sites, binding sites on DNA, e.g. the positions of nucleosomes on a minichromosome. The DNA protected by binding proteins is cleaved in susceptible, i.e. unprotected, regions by non-specific chemical or enzymic methods, the protecting protein is removed, and a restriction nuclease introduces reference points recognised by complementary polynucleotide probes for subsequent analysis of the sequences from the restriction sites towards the sites at which it was initially cleaved in the unprotected regions.

■ **induced fit**

a change in the conformation of an enzyme in response to substrate binding that renders the enzyme catalytically active. It is one of the models used to explain substrate specificity.

■ **inducer**

a signal molecule that, when bound to a regulatory protein, produces an increase in the expression of a given gene.

■ **inducible proteins**

those which are synthesised in different amounts depending on cellular signals.

■ **induction**

an increase in the expression of a gene in response to a change in the activity of a regulatory protein.

■ **informational macromolecules**

biomolecules containing information in the form of specific sequences of different monomers; for example, many proteins, lipids, polysaccharides, and nucleic acids.

■ **infra-**

below, underneath.

■ **infrared spectroscopy**

IR spectroscopy; a technique that measures absorption of electromagnetic radiation between the long wavelength end of the visible spectrum (780-800 nm) and the short end of the microwave spectrum (1000 m). In practice, most commercial IR spectrometers measure from 780 nm to 300 μm. Absorption in the IR is due to interatomic motions of molecules, vibrations and rotations about bonds. Spectra are commonly presented as percent transmission vs. wavelength.

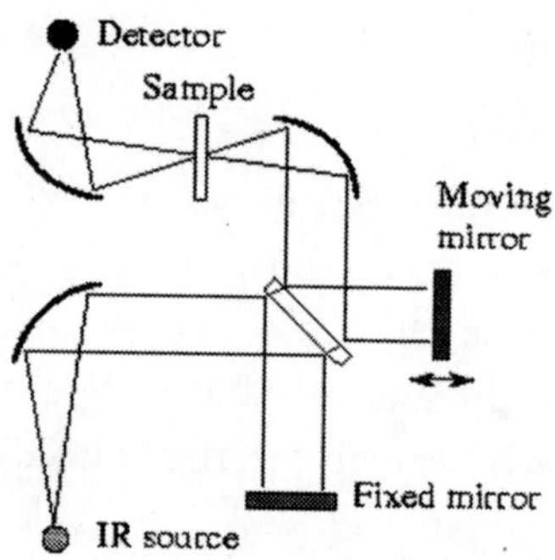

■ **initiation codon**

AUG. Codes for the first amino acid in all polypeptide sequences, which is N-formyl-methionine in prokaryotes and methionine in eukaryotes.

■ **initiation complex**

a complex of all ribosomal components, a mRNA and the initiating Met-tRNA^{Met} (eukaryotes) or fMet-tRNA^{Met} (prokaryotes).

■ **initiation factor**

an accessory protein that is necessary for assembly of the ribosome-mRNA complex and the start of protein synthesis.

■ **inorganic pyrophosphatase (pyrophosphatase)**

an enzyme that hydrolyses a molecule of inorganic pyrophosphate to yield two molecules of phosphate. reaction.

■ **insertion mutation**

a mutation caused by insertion of one or more extra bases between bases in DNA, or a mutagen which itself is inserted between bases (like ethidium bromide).

■ **insertion sequence**

the specific base sequences at either end of a transposon which allows for insertion into DNA.

■ **insulator boundary element**

a sequence in genomic DNA which sets a limit to heterochromatin structure.

■ **integral membrane proteins**

proteins inserted into a membrane by hydrophobic interactions Contrast with peripheral proteins.

■ **inter-**

among, between.

■ **intercalating agent**

a chemical, usually containing aromatic rings, that can sandwich in-between adjacent base pairs in a DNA duplex. The intercalation leads to an adjustment in the DNA secondary structure, as adjacent base pairs are usually close-packed.

■ **intercalating mutagen**

a mutagen that inserts itself between two successive bases in a nucleic acid causing a frame-shift mutation.

■ **intercalation**

insertion between two stacked aromatic or planar rings. Most commonly refers to the insertion of a planar mutagen between two successive bases in a nucleic acid.

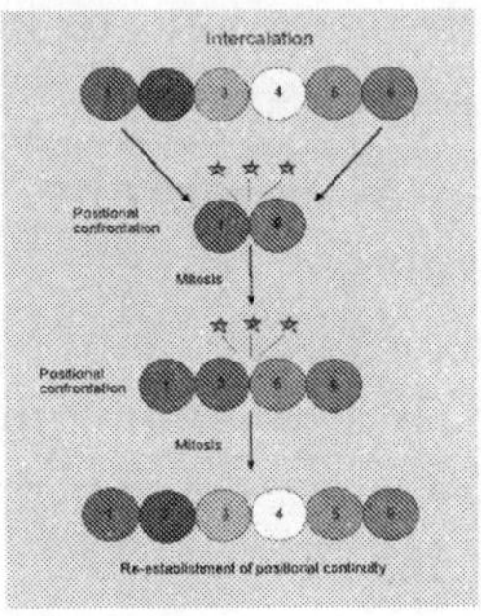

■ **interferon**

one of a family of proteins that are liberated by special host cells in the mammal in response to viral infection. The interferons attach to an infected cell, where they stimulate antiviral protein synthesis.

■ **intermediary metabolism**

the enzyme-catalysed reactions that extract chemical energy from nutrient molecules and utilise it to synthesise perform cell functions.

■ **internal guide sequence**

a polynucleotide sequence near the 5'-end of group I introns that pairs with sequences of the upstream exon in an intermediate of the self-splicing process.

■ **intervening sequence**

see intron

■ **intro-**

inward, during.

■ **intron**

a segment of the nascent transcript that is removed by splicing. Also refers to the corresponding region in the DNA. Synonymous with intervening sequence.

■ **invertasome**

a nucleoprotein complex that mediates inversion of a DNA sequence, bracketed by two recombination sites, within the genome of an organism.

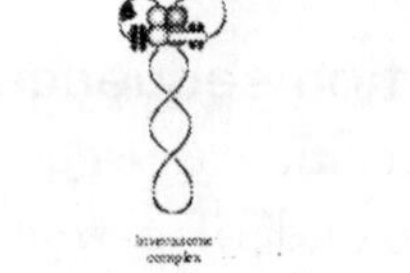

■ **inverted repeat**

a chromosome segment that is identical to another segment on the same chromosome except that it is oriented in the opposite direction.

■ ion channel

an integral membrane protein that provides for the regulated transport of (a) specific ion(s) across a membrane.

■ ion product of water (Kw)

the product of the concentrations of H^+ and OH^-. For pure water $Kw = [H^+][OH^-] = 1 \times 10^{-14}$ at 25 °C.

■ ion trap

a mass spectrograph in which vaporised ions are confined by an AC electric field. As the radio-frequency voltage that confines the ion is changed, the ion will escape the trap at a voltage characteristic of its mass-to-charge ratio, and be detected. Sequences of polypeptides and polynucleotides may be determined by expulsion from the trap of all but one kind of large mass ion, fragmentation of this ion by collision with helium atoms and analysis of the mass spectrum of the resulting fragments. A variant of this electrical-type ion trap uses a magnetic field to confine ions.

■ ion-exchange resin

a polymeric resinous substance, usually in bead form, that contains fixed groups with positive or negative charge. An anion exchange resin has positively-charged groups and is therefore useful in exchanging the anionic groups in a test sample; a cation exchange resin is itself negatively charged, and has the opposite application.The resin is usually used in a column chromatographic procedure.

■ ionising radiation

any radiation that causes loss or gain of electrons from molecules, thus making them more reactive.

■ ionophore

a compound that binds metal ions and diffuses across membranes carrying the bound ion. An example is valinomycin, which transports potassium.

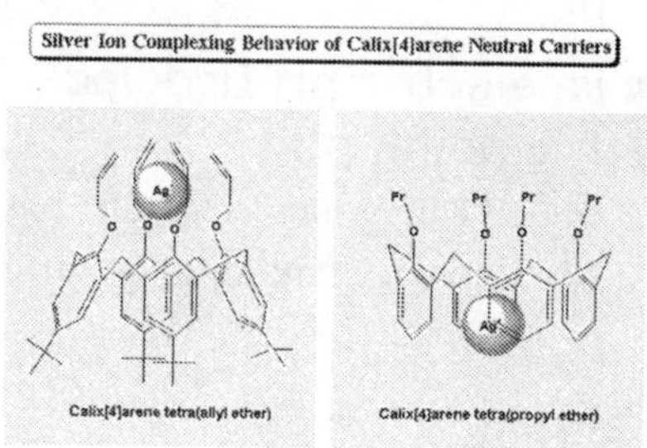

■ **iontophoresis**

the transport of ions across a semi-permeable membrane under the influence of an electric charge. As ions are driven across the membrane by the electric field, there is also the accompanying transport of bulk water (electro-osmosis) and uncharged solutes.

■ **iron-sulphur center**

a prosthetic group of some redox proteins involved in respiration and other electron transfers. Fe^{2+} or Fe^{3+} is with Cys groups in the protein.

■ **ism**

theory, characteristic of.

■ **iso-**

equal, same.

■ **isoelectric focusing**

an electrophoretic method for separating macromolecules on the basis of their pI.

■ **isoelectric pH (isoelectric point, pI)**

the pH at which a solute has no net electric charge and thus does not move in an electric field.

■ **isoelectric point or pH**

the pH at which a protein has no net charge.

■ **isoenzymes**

isozymes.

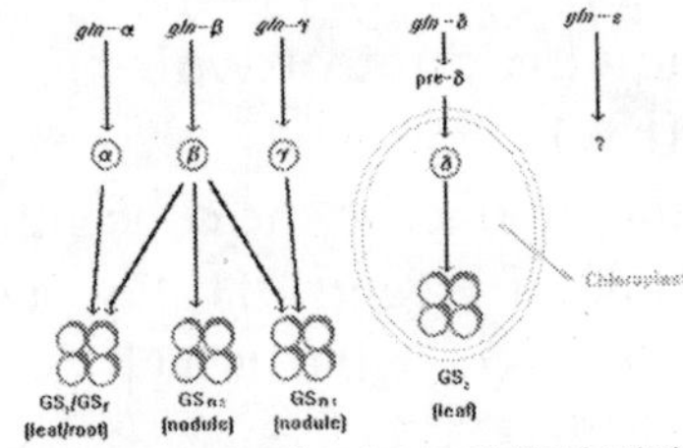

Genetic control of glutamine synthetase isoenzymes in *Phaseolus vulgaris*

■ **isomerases**

enzymes that catalyse the transformation of compounds into their positional isomers.

■ **isomerisation**

rearrangement of atomic groups within the same molecule without any loss or gain of atoms.

■ **isomers**

any two molecules with the same molecular formula but a different arrangement of molecular groups: either stereospecific or geometric.

■ **isoprene**
the hydrocarbon 2-methyl-1,3-butadiene, a recurring structural unit of many lipid molecules.

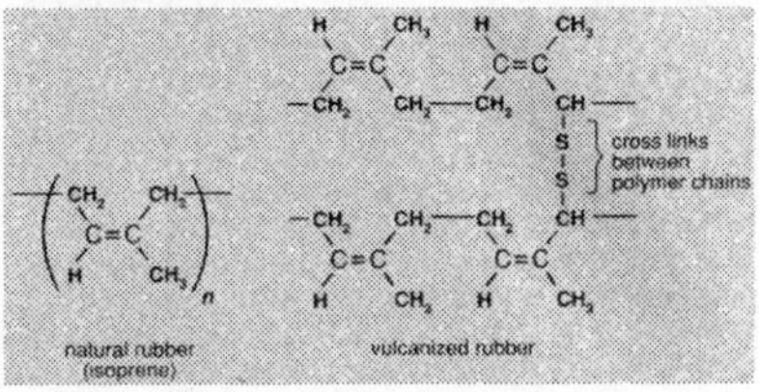

■ **isoreceptor**
one of two or more homologous plasma membrane or cytosolic receptors that have the same or altered functions; analogous to an isoenzyme.

■ **isothermal**
occurring at constant temperature.

■ **isotope effect**
the effect on kinetics of the substitution of one isotope for another in a bond that must be broken in a rate-limiting step (primary isotope effect); especially prominent in the case of cleavage of a bond to a deuterium atom compared with a hydrogen atom. Smaller effects are seen in the case of a substitution of an isotope for a non-reacting atom (secondary isotope effect); effects are also seen on parameters other than the rate-limiting step.

■ **isotopes**
stable or radioactive forms of an element that differ in atomic weight but are otherwise chemically identical to the naturally abundant form of the element. These properties allow them to be used as tracer molecules.

■ **isozymes**
two or more forms of an enzyme that catalyse the same reaction but differ from each other in their physical properties, such as amino acid sequence, substrate affinity, Vmax, tissue expression or regulatory properties.

■ **-itis**
inflammation.

■ **-ity**
makes a noun of quality.

■ **-ium**
thing.

■ **-ize**
do.

■ **J-chain**
a polypeptide found in the dimeric immunoglobulin A and the pentameric immunoglobulin M, involved in joining together the subunits of each multimer.

■ **jumping library**
a tool for the mapping and sequencing of large stretches of a chromosome. Starting at one point in the polynucleotide sequence, a distal point is located without having to first sequence the intervening bases. The double-stranded DNA is divided into large fragments with a rare cutter and one end of each piece is ligated to a selective marker and cyclised, then cut with a frequent cutter. The marker allows isolation of the two ends of the original large piece without the intervening sequence.

■ **jumping PCR**
a technique for reconstruction and replication of randomly damaged double-stranded DNA, e.g. that found in mummified tissues or museum samples. In the first PCR cycle each primer is extended until a break in the template is encountered, but in the second cycle the newly synthesised strands find other templates that are intact in the region where the first templates were damaged. Each cycle extends the length of newly synthesised strands until complete strands are formed and PCR can begin its exponential synthesis.

■ **junk DNA**
dNA for which a function has yet to be identified. An elaborate nomenclature has been proposed for this DNA: any stretch of DNA is a nuon; DNA with an evolutionary potential for development of function is a potonuon; DNA that was previously non-functional, or that had a completely different function and has been evolutionarily co-opted for a new function (a process termed exaptation), is a xaptonuon; and a non-co-opted potonuon is a naptonuon.

■ **juxta-**
adjacent to.

■ **karyokinesis**
division of the nucleus of a cell during mitosis.

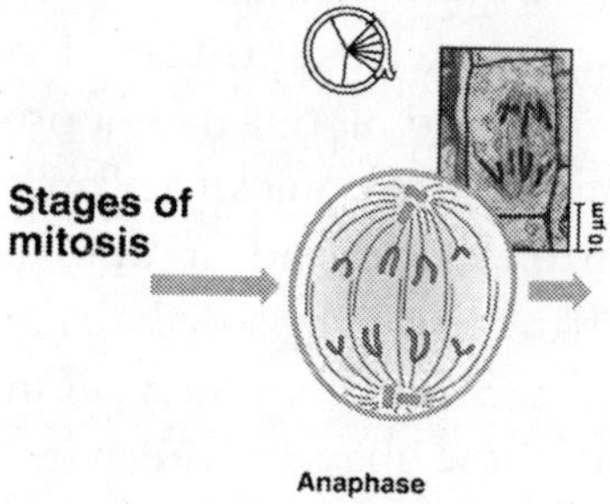

■ **karyophile**
a substance that can diffuse into and be bound in the nucleus of a cell, presumably due to its affinity for a non-diffusible nuclear compound.

■ **karyotype**
the display of chromosomes, arranged from largest to smallest, obtained by cutting out and arranging photomicrographs of chromosomes.

■ **katal**
a measure of enzyme activity; the conversion of 1 mol of substrate per s under specified conditions.

■ **keratins**
insoluble protective or structural proteins consisting of parallel polypeptide chains in α-helical or ß conformations.

■ **ketogenic**
describing amino acids that are metabolised to acetoacetate and acetate.

■ **ketogenic amino acid**
an amino acid whose carbon skeleton can be converted, at least in part, into ketone bodies.

■ **ketone**
a functional group of an organic compound in which a carbon atom is double-bonded to an oxygen. Neither of the other substituents attached to the carbon is a hydrogen. Otherwise the group would be called an aldehyde.

■ **ketone bodies**
a group of molecules normally produced in very low amounts, but increased amounts are found during fasting or in untreated diabetes mellitus. The three ketone bodies are acetoacetate, D-ß-hydroxybutyrate, and acetone.

■ ketose

a simple monosaccharide in which the carbonyl group is a ketone.

■ ketosis

an accumulation in the blood of ketone bodies, indicative of a metabolic dysfunction such as uncontrolled diabetes mellitus or starvation.

■ kilobase (kb)

a measure of DNA chain length; 1000 nucleotide residues.

■ kinases

enzymes that catalyse the phosphorylation of certain molecules using ATP as a phosphate source.

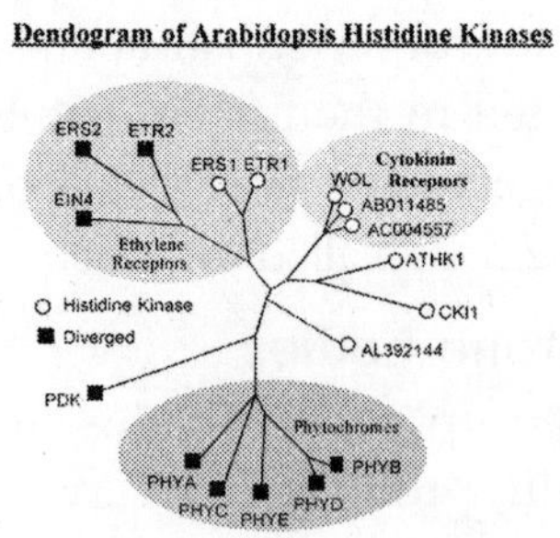

■ kinetic assay

an enzyme-based assay that measures the amount of substrate present by correlation of the rate of reaction with the known dependence of the rate on substrate concentration, usually under first-order conditions.

■ kinetic partitioning

the hypothesis that in folding of a newly synthesised or renaturing biopolymer, a protein or RNA quickly collapses into condensed structures characterised by a high degree of secondary structure. One of these may be the native conformation (the others are "kinetic traps"), but significant energy barriers isolate these forms. Kinetic partitioning is the hypothetical basis of recognition of proteins by molecular chaperones: quickly after synthesis, a cytosolic protein achieves its native conformation and thus avoids binding to the chaperone, while other proteins, e.g. those with leader sequences, much more slowly fold into native structures, which affords the chaperone the opportunity to bind them.

■ kinetics

the study of rates of reactions.

■ **kinetochore**

the site on a metaphase chromosome to which chromosomal microtubules attach and in anaphase draw them toward the centrioles.

■ **kinin**

a hypotensive plasma peptide that dilates small blood vessels and increases capillary permeability, e.g. bradykinin; formed from its precursor, a kininogen, by the proteolytic action of a kallikrein.

■ **kininogen**

a protein that contains the sequence Arg-Pro-Pro-Gly-Phe-Ser-Pro-Arg which has inhibitory activity against cysteine proteinases. A kininogen is a precursor of a kinin, which is a decapeptide or undecapeptide produced by the proteolytic activity on a kininogen by a kallikrein. An L-kininogen is a low-molecular-mass kininogen and an H-kininogen is a high-molecular-mass kininogen, e.g. T-kininogen is one of the H-kininogens in the rat.

■ **kissing complex**

the interaction of two RNA molecules by hybridisation of complementary sequences from a loop of each.

■ **Klenow fragment**

the product of limited proteolysis of E. coli DNA polymerase I that retains the polymerase and 3'-to-5' exonuclease (proofreading) activity but is missing the 5'-to-3' nuclease activity.

■ **Km**

see Michaelis constant

■ **knock-down**

a strategy for down-regulation of expression of a gene by incorporation into the genome an antisense oligodeoxynucleotide or ribozyme sequence that is directed against the targeted gene.

■ **knock-in**

an organism in which a foreign or mutated gene replaces the normal gene. Unlike in transgenic techniques, the insertion is site-specific and delivers one copy of the gene.

■ **knock-out**

an animal or plant from which a certain gene has been deliberately deleted, in order to observe, as a test for the gene's physiological significance, the viability or phenotype of the organism. A knock-out is contrasted to a transgenic, in which a gene from another genomic source has been inserted.

■ **Krebs cycle**

a cyclic system of enzymatic reactions for the oxidation of acetyl residues (as acetyl CoA) to carbon dioxide. The formation of citrate is the first step. Also called citric acid cycle or tricarboxylic acid (TCA) cycle.

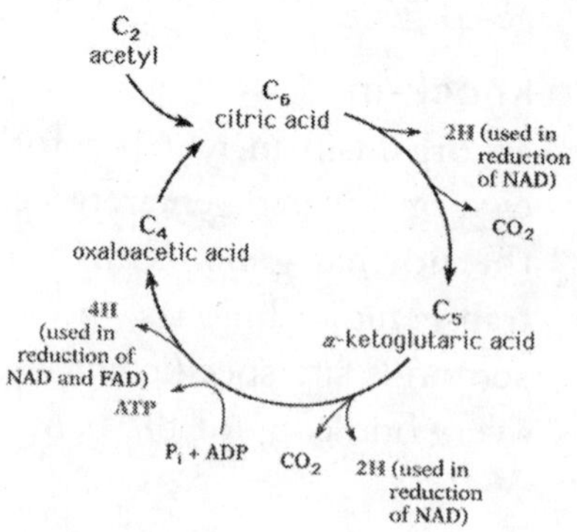

■ **kringle**

a protein folding motif, as visualised in two dimensions, of loops formed by multiply disulphide-bridged sequences of several circulating proteins, e.g.alpha 2-macroglobulin, complement component C3, prothrombin, tissue-type plasminogen activator.

■ **label**

a chemical modification or isotope substitution that makes possible the identification of a molecule, moiety or atom throughout its metabolic transformations or transport.

■ **labile sulphide**

a compound from which H_2S evolves by treatment with acid, e.g. the sulphur of Fe-S proteins.

■ **lability**

instability to heat, shear or other physical or chemical stress.

■ **lagging strand**

the DNA strand that, during replication, must be synthesised in the direction opposite to that in which the replication fork moves. Synthesised *via* a series of fragments called Okazaki fragments.

■ lamellipodium (ruffled edge)

a fin-like projection of the plasma membrane of a eukaryotic cell.

■ lampbrush chromosome

a chromosome in which the chromatin extends out from a central axis into many loops, where transcription and other activity occurs.

Lampbrush chromosome

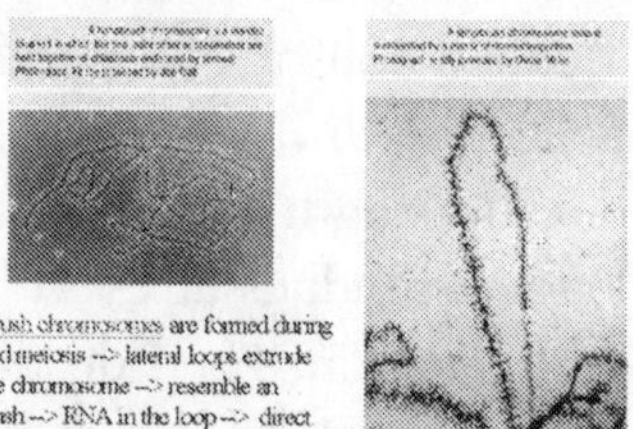

Lampbrush chromosomes are formed during extended meiosis --> lateral loops extrude from the chromosome --> resemble an lampbrush --> RNA in the loop --> direct observation of gene expression in native state.

■ lampbrush chromosome

giant diplotene chromosome found in the oocyte nucleus. The loops that are observed are the sites of extensive gene expression.

■ lathyrism

a skeletal malformation due to dietary beta-aminopropionitrile (present in the seeds of sweet peas, Lathyrus odoratus) which interferes with the proper cross-linking of collagen.

■ law of mass action

the law stating that the rate of any given chemical reaction is proportional to the product of the concentrations of the reactants. At extremely low concentrations, activities are used.

■ lawn

the thick coverage of an agar (or other gel medium) plate with bacteria; usually as a test system for a lysogenic bacteriophage.

■ leader

a short sequence near the amino terminus of a protein or the 5' end of an RNA molecule that has a specialised targeting or regulatory function. Often the leader is cleaved after the targeting event.

■ leader region

the region of an mRNA between the 5' end and the initiation codon for translation of the first polypeptide chain.

leader sequence

signal sequence; the N-terminal portion of a secretory or membrane protein that assists it across the membrane of the rough endoplasmic reticulum, where it is synthesised, but which is cleaved from the protein even before the synthesis of the protein is complete. Mitochondrial and chloroplast proteins that are synthesised on ribosomes also have signal sequences that target them to their organelle; these may be N-terminal or internal, and may or may not be removed by proteolysis.

leading strand

the DNA strand that, during replication, is synthesised in the same direction in which the replication fork moves. Synthesised in a continuous fashion.

leaflet

one half of a lipid bilayer.

lectins

agglutinating proteins usually extracted from plants.

Leloir pathway

the metabolic route by which galactosyl and glucosyl units are interconverted; the transformation of galactose and ATP to glucose-1-phosphate via the galactokinase, galactose-1-phosphate uridyltransferase and UDP-galactose-4-epimerase reactions.

lethal mutation

a mutation that inactivates a biological function essential to the life of the cell or organism.

leucine scissors

a supersecondary structure of some DNA-binding proteins; a coiled coil of two parallel -helices, each of about 30 residues in length and containing leucine residues at every seventh position. The side chains interact with non-polar side chains of the other helix to hold the two together by hydrophobic forces.

leucine zipper

a protein structural motif involved in protein-protein interactions in many eukaryotic regulatory proteins (C/EBP prototype). Contain a repeat structure: Leu residues in every seventh position, causes a large amount of DNA to loop out.

■ leukotriene

a family of molecules derived from arachidonate having in common 3 conjugated double bonds. In general they cause muscle contractions that constrict air passages in the lungs, involved in asthma.

■ levorotatory isomer

a stereoisomer that rotates the plane of plane-polarised light counterclockwise.

■ ligand

a (usually small) molecule that binds to another, such as oxygen when it binds to myoglobin.

■ Ligase

an enzyme that catalyses the joining of two molecules together. In DNA it joins 5'-OH to 3' phosphates.

■ ligase

one of a class of enzymes that join two substrate molecules in energy- (usually ATP-) dependent reaction, e.g. an amino acyl-tRNA synthetase, a carboxylase; in molecular biology, an enzyme that attaches the 3'-end of one polynucleotide to the 5'-end of another.

■ ligase chain reaction (LCR)

a technique for detection and amplification of target DNA sequences. Two oligodeoxynucleotides are synthesised which between them are complementary to the entire target sequence, one to the 5'-side and one to the 3'-side. If the target sequence is present in the DNA sample under examination, the oligonucleotides will bind to it at moderate temperature with their ends abutting in the centre, and a heat-stable ligase will join them into a complete polynucleotide. No ligation will occur if the target sequence is absent or if the match between synthetic oligonucleotides and target sequence is imperfect in the region where they abut. At an elevated temperature the new polynucleotide dissociates from the original DNA template, and upon cooling to moderate temperature it and the original DNA serve as templates for a second cycle of hybridisation, ligation and thermal dissociation. At each cycle there is a doubling of the number of new complete polynucleotides, so after 20 cycles there can be a 106-fold amplification. The half-sequences may be constructed with markers on the non-ligating ends to assist isolation (e.g. biotin) or detection (e.g. a fluorophore).

■ light reactions
the reactions of photosynthesis occuring only in the light.

■ light subunit
the smaller of the two ribonucleoprotein complexes that make up a ribosome; more generally, the smallest of the subunits of any complex.

■ Lineweaver-Burk equation
an algebraic transformation of the Michaelis-Menten equation (plot of 1/V *vs.* 1/[S]), allowing determination of Vmax and Km by extrapolation of [S] to infinity.

■ linkage
the tendency of markers to be inherited together.Linkage of two markers is an indication that they are close to one another in the genome.

■ linkage group
a collection of genes that are inferred to be located together on a single chromosome because of the pattern of their inheritance.

■ linkers
short oligonucleotides that can be ligated (connected) to larger DNA fragments, then cleaved (cut) to yield overlapping cohesive (sticky) ends, suitable for ligation to other DNAs that contain comparable cohesive ends.

■ linking clone
a tool for the localisation of restriction fragments within a physical map of the genome of an organism. A cloned fragment that contains very few of an already rare restriction site is used as a hybridisation probe for chromosomal DNA digested with the rare cutter.

■ linking number
the net number of times one polynucleotide chain crosses over another polynucleotide chain.By convention, right-handed crossovers are given a plus designation.

lipases
enzymes that catalyse the hydrolysis of triacylglycerols.

lipid
a biological molecule that is soluble in organic solvents.Lipids include steroids, fatty acids, prostaglandins, terpenes, and waxes.

lipid bilayer
model for the structure of the cell membrane based on the interaction between the hydrophobic regions of phospholipids.

lipoate (lipoic acid)
a coenzyme involved as a carrier of acyl groups.

lipocalin
one of a family of homologous proteins that bind lipophilic substances, including retinol-binding protein,beta-lactoglobulin, orosomucoid.

lipopolysaccharide
usually refers to a unique glycolipid found in Gram negative bacteria.

lipoprotein
a protein-lipid complex that functions in the transport of lipids in the blood. In cyclical fashion, present as chylomicrons, VLDL, LDL, IDL, HDL or chylomicron remnants. The protein component alone is an apolipoprotein.

liposome
a synthetic micelle formed by dispersion of a polar lipid in aqueous solution in such a way that it forms into lipid-bilayer-encapsulated vesicles; useful for packaging of drugs or enzymes for introduction into cells.

locus
the position on a chromosome of a particular allele.

logy
study of, reasoning about.

low-energy phosphate compound
a phosphorylated compound with a relatively small standard free energy of hydrolysis.

Lowry protein assay
a refinement of the colorimetric method for protein determination that uses the Folin-Ciocalteu phosphomolybdotungstate reagent.

■ **luciferin**

a potentially chemiluminescent substrate; the enzymic reaction of an enzyme (a luciferase) converts it to an excited state that decays with emission of visible radiation; e.g. the benzothiazole of fireflies which upon adenylation is decarboxylated and deadenylated with the emission of yellow light. The energy of photon emission derives from an oxidation with molecular oxygen that may be the enzymic or a non-enzymic reaction of the product of the enzymic reaction. There is no structural homology within the luciferins or luciferases, which have a wide biological distribution.

■ **lyases**

enzymes that catalyse the removal of a group from a molecule to form a double bond, or the addition of a group to a double bond.

■ **lymphocytes**

a subclass of leukocytes characterised as mononuclear, with a small amount of cytoplasm. Involved in the immune response and divided into two types: B lymphocytes (synthesise and secrete antibodies) and T lymphocytes (participate in cell-mediated immunity).

■ **lymphokine**

a protein secreted by T-cells that functions in cell-mediated immunity to activate or inhibit leucocytes.

■ **lymphoma**

a cancer of lymphatic tissue.

■ **lysate**

the product of lysis; a broken-cell preparation.

■ **lysis**

destruction of a cell's plasma membrane or of a bacterial cell wall, releasing the cells contents and killing the cell.

■ **lysogenic bacterium**

a bacterium that hosts a prophage, which may lie dormant even for many generations but eventually may cause lysis.

■ **lysogenic virus**

a virus that can adopt an inactive (Iysogenic) state, in which it maintains its genome within a cell instead of entering the Iytic cycle. The circumstances that determine whether a Iysogenic (temperate) virus will adopt an inactive state or an active Iytic state are often subtle and depend upon the physiologic state of the infected cell.

■ lysosome

a membrane-bounded organelle in the cytoplasm of eukaryotic cells whose function is to degrade and recycle proteins, in particular.

■ lytic infection

a virus infection that leads to the Iysis of the host cell, yielding progeny virus particles.

■ M phase

that period of the cell cycle when mitosis takes place.

■ macro-

large.

■ macroenzyme

a normal, active enzyme that appears in serum as a conjugate, often with a specific autoantibody, thus giving it a greatly increased molecular weight.

■ macrolide

a moiety of some antibiotics (e.g. erythromycin, methymycin) which consists of a large lactone ring, synthesised largely or entirely from propionate-derived units.

■ macromolecule

a molecule having a molecular weight in the range of a few thousand to many millions: proteins, nucleic acids and polysaccharides.

■ macrophage

one kind of phagocytic cell.

■ major groove

the wider of the two helical spaces on the surface of an A- or B-DNA double helix. The other helical space is the minor groove.

■ manganese centre

in the photosynthetic photosystem II, a complex of four manganese atoms separated by oxygen atoms that accumulates a positive charge that is used to oxidise water to molecular oxygen.

■ Manton-Gaulin

homogeniser
an apparatus for disruption of cells, especially the large quantity of bacteria or yeast that result from a fermentation, by pumping a paste of the collected cells through an orifice of adjustable size in the head of a reciprocating piston so that the shearing forces that develop break apart the cells.

■ **marker**
a "landmark" that can be localised to a specific region of the genome.

■ **marker rescue**
a technique to localise the site of a mutation within a gene. A polynucleotide, the 'marker', from the known wild-type sequence is allowed to recombine with the mutant genome such that, if it includes the site of the mutation, it 'rescues' the mutant by causing it to revert.

■ **mass-action ratio**
the ratio which describes the kinetics of a reaction, given as products over reactants. For a hypothetical reaction aA + bB → cC + dD, the ratio would be$[A]^a [B]^b/[C]^c [D]^d$.

■ **mating type**
in effect, a sex of a haploid yeast cell that restricts it to forming a diploid cell only with a haploid of a different mating type.

■ **medi-**
middle.

■ **medium**
the solid or liquid substratum on which, or in which, cells or organ explants can be made to grow; a medium may include well defined factors such as salts, amino acids and sugars, as well as less well defined factors such as serum or blood.

■ **medulla**
soft inner part.

■ **mega-**
large.

■ **megadose**
an amount of a dietary supplement, such as a vitamin, mineral or essential amino acid, that is orders of magnitude greater than the officially recommended dosage.

megalo-
very large.

megaly
large.

meiosis
a type of cell division in which diploid cells give rise to haploid cells destined to become gametes.

membrane
a sheet-like composite of protein and lipid that is the boundary of cells and organelles.

membrane potential
a charge difference across a membrane due to the Donnan effect.

membrane protein
a protein that is associated with a membrane, rather than found free in the cell.A membrane protein may be integral (embedded or buried) in the membrane, or peripheral (attached more loosely, by interactions with either lipid or intergral membrane proteins).

membrane skeleton
the matrix of proteins and fibres that underlies the lipid bilayer of a plasma membrane; it contains actin filaments, actin-binding proteins and/or cell-specific structural proteins.

membrane transport
movement of a polar molecule across a membrane *via* a specific membrane protein (a transporter).

merodiploid
an organism that is diploid for some but not all of its genes.

Meselson-Radding model
a conceptualisation of general genetic recombination during meiosis of newly replicated double-stranded DNA. An endonuclease nicks one strand of one daughter duplex; the new 5'-end dissociates and allows synthesis at the new 3'-end to permanently displace it. The free 5'-end displaces its homologue in the other daughter duplex, followed by excision of that displaced loop and ligation of the first 5'-end to the new 3'-end of the second duplex. Ligation of the remaining 3'- and 5'-ends and rotation to exchange the intact strands

allows branch migration and separation, as in the Holliday model.

Meselson-Stahl experiment
a demonstration of semi-conservative DNA replication. E. coli grown on ^{15}N salts for several generations produce labelled DNA that is more dense than normal DNA, as judged by density gradient equilibrium centrifugation. When transferred to 14N media, the first generation produces double-stranded DNA all of intermediate density, and the second generation DNA half of which is of intermediate density and half of normal density.

meso-
middle.

mesoderm
one of the three primordial germ layers formed during early embryogenesis; a precursor of muscle, adipose tissue, blood vessels, the gastrointestinal tract, etc.

mesosome
an invagination of the bacterial cell membrane.

messenger RNA (mRNA)
a class of RNA molecules, each of which is complementary to one strand of DNA, characterised by a 5' 7-N-methylguanine cap and a 3' polyA tail. Its function is to carry the genetic message from the chromosome to the ribosomes.

meta-
beyond, between.

metabolic turnover
a measure of the rate at which already existing molecules of the given species are replaced by newly-synthesised molecules of the same type.Usually isotopic labeling is required to measure turnover.

metabolism
the sequences by which foodstuffs are degraded for the energy that is released and for transformation into cellular components.

metabolite
a chemical intermediate in the enzyme-catalysed reactions of metabolism.

■ **metalloprotein**

a protein having a metal ion as its prosthetic group.

■ **metamorphosis**

a change of form, especially the conversion of a larval form to an adult form.

■ **metaphase**

that stage in mitosis or meiosis when all of the chromosomes are lined up on the equator (i.e., an imaginary line that bisects the cell).

■ **methyl trap**

the accumulation of 5-methyltetrahydrofolate in vitamin B_{12} deficiency or related conditions (e.g. pernicious anaemia) due to the inability to transfer the methyl group to homocysteine by the B_{12} coenzyme-dependent transferase.

■ **micelle**

a very small aggregate of matter that is dispersed in solution; often of lipoid material stabilised by detergents, or even of detergent only.

■ **Michaelis-Menten constant (Km)**

practically speaking, the substrate concentration at which an enzyme-catalysed reaction proceeds at one-half its maximum velocity. Rigorously, the ratio $k_{-1} \bullet k_2/k_1$.

■ **Michaelis-Menten equation**

the equation describing the hyperbolic dependence of the initial reaction velocity, Vo, on substrate concentration, [S], in many enzyme-catalysed reactions: $Vo=Vmax[S]/Km+[S]$

■ **Michaelis-Menten kinetics**

a kinetic pattern observed in enzyme-catalysed reactions when the initial rate (as a function of substrate concentration) exhibits a hyperbolic shape.

■ **micro-**

small.

■ **microcluster**

an aggregate of several (2-10) membrane receptors that forms in response to cell stimulation; such an aggregate is a transitory state between the receptors being dispersed in the membrane and their formation into much larger aggregates, patches that will undergo endocytosis.

■ microfilaments

thin filaments composed of actin found throughout the cytoplasm of eukaryotic cells. Their function is give the cell shape and movement.

■ micron

a micrometer (m) / 10^{-6} m.

■ microsomes

membranous vesicles formed by fragmentation of the endoplasmic reticulum (and probably Golgi) of eukaryotic cells; recovered by differential centrifugation.

■ microtubule

a large (30nm diameter) rigid component of the cytoskeleton that is built of alpha and beta-tubulin subunits and associated proteins and serves as a component of cilia, eukaryotic flagella and mitotic spindles.

■ mimotope

a synthetic peptide that binds to an antibody against an unidentified antigen; usually a peptide selected from a large library of peptides.

■ minisequencing

a multiplex method for scanning genomic DNA samples for the presence of specific mutations. PCR primers complementary to the mutant and wild-type sequences are immobilised at coded loci on a surface and exposed to genomic DNA samples. Incorporation of a fluorescent marker during PCR amplification will indicate by the presence of fluorescence at a locus, the existence there of the mutant or wild-type template in the genomic DNA to which it had been exposed.

■ mismatch repair

the replacement of a base in a heteroduplex structure by one that forms a Watson-Crick base pair.

■ mismatch repair analysis

an assay system for detection of single-base mismatches in a DNA fragment of up to 10 Kb in length, which depends upon an Escherichia coli repair system that detects mismatched bases between a newly synthesised (unmethylated) DNA strand

and an unreplicated (methylated) strand. The unmethylated DNA strand in a hybrid of the two is nicked, digested and resynthesised on the methylated template. This system, however, does not replace insertions or loops larger than 5 bases. A mutated DNA fragment is inserted into a plasmid that contains a reporter gene, such as LacZ, which can be used to identify colonies that express the plasmid, and is then transformed into an E. coli dam- strain, which results in an unmethylated plasmid. The wild-type DNA fragment is inserted into the same plasmid in which a 5-base sequence interrupts the reporter gene; this is transformed into a dam+ strain, which fully methylates the plasmid. The two strains are grown, the plasmids isolated, mixed, denatured and reannealed. Fully methylated homoduplexes are digested with DpnI and unmethylated homoduplexes are digested with MboI. Surviving heteroduplexes, which contain both the intact and the interrupted reporter genes, are circularised, ligated and transformed into E. coli. If there is no mismatch, both strands of the plasmid will be replicated and the intact reporter gene will be expressed. However, if there is a mismatch, the non-methylated strand, along with the intact reporter gene will be digested and will not be expressed.

■ **missense mutation**

a change in which a codon for one amino acid is replaced by a codon for another amino acid.

■ **mitochondrial matrix**

the aqueous portion of a mitochondrion bounded by the inner mitochondrial membrane.

■ **mitochondrion**

membrane-bounded organelle in the cytoplasm of eukaryotes sometimes referred to as powerhouse of the cell because it contains the enzyme systems required for the citric acid cycle, fatty acid oxidation, electron transfer chain, and oxidative phosphorylation(plural, mitochondria).

■ **mitosis**

the multistep process (interphase, prophase, metaphase, anaphase, telophase) in eukaryotic cells that results in the replication of chromosomes and cell division.

■ **mixed-function oxidases (oxygenases)**

enzymes that use O_2 to simultaneously oxidise a substrate and a cosubstrate (commonly NADH or NADPH).

■ **mobile barrier**

a proposed mechanism of membrane transport. A transmembrane protein, the activated porter, exposes its binding site for the transported molecule to one membrane face and then to the other. In the case of a uniport porter, activation does not depend upon the solute, so transport can be effected in either direction; a symport porter is activated only when both (or neither) of the two solutes are bound; an antiport porter is activated when either, but not both, of the two solutes is bound. This proposal is often coupled with the concept of solvation substitution, the identification of solute binding as the replacement of bound water by specific liganding groups of the porter, e.g. chelation of ions, to help explain the specificity of the porter and how it maintains a barrier against transport of water across the membrane.

■ **mobile genetic element**

a segment of the genome that can move as a unit from one location on the genome to another, without any requirement for sequence homology.

■ **modulator**

a metabolite that, when bound to the allosteric site of an enzyme, alters its kinetic characteristics. Can be homotropic or heterotropic, inhibitor or activator.

■ **mole**

one gram molecular weight of a compound, containing 6.02 x 10^{23} molecules.

■ **molecular channel**

a proposed feature of multienzyme complexes; the product of catalysis by the first enzyme of the metabolic sequence travels through the channel to the active site of the next enzyme, etc.

molecular dynamics
the computer-simulated quantitative evaluation of a molecular model by assignment of force constants to bonded and non-bonded interactions between neighbouring atoms so that bond distances and angles are confined to an acceptable range and crowding is minimised; often coupled with energy minimisation, which makes small adjustments in the locations of atoms to optimise these interactions.

molecular organelle
a very large, non-membrane-limited aggregate of subunits that in prokaryotes functions like an organelle. They occur also in eukaryotes, e.g. the proteasome, which as a 26 S multimer is a cylinder 40 micrometer long and 20 micrometer in diameter.

molecularity of a reaction
the number of molecules involved in a specific reaction step.

monoclonal antibodies
antibodies produced by a cloned hybridoma cell, which therefore are identical and directed against the same epitope of an antigen.

monocyte
one type of non-granular leucocyte; a precursor of macrophages and osteoclasts.

monolayer
a single layer of oriented lipid molecules.

monolayer
a single layer of oriented lipid molecules or cells.

monomer
one unit of a protein or other structure.

monoprotic acid
an acid having only one dissociable proton.

monosaccharide
a carbohydrate consisting of a single sugar unit.

motility assay
a method for detection of a molecular motor which depends upon observation of nanoscale motion; e.g. the detection of the rotary motion of a fluorophore attached via an actin arm to the gama -subunit rotating inside the alpha - and beta-subunit hexamer of an immobilised ATP synthase.

mRNA
messenger RNA.

mucopolysaccharide
an older name for a glycosaminoglycan.

multienzyme system
a group of related enzymes participating in a given metabolic pathway and often channeling substrates from one to another.

multigene family
a group of genes that code for homologous products with similar functions, e.g. the genes for globins.

mutagen
an agent that can bring about a heritable change (mutation) in an organism.

mutagenesis
a process that leads to a change in the genetic material that is inherited in later generations.

mutant
an organism that carries an altered gene or change in its genome.

mutarotation
the change in optical rotation of a sugar that is observed immediately after it is dissolved in aqueous solution, as the result of the slow approach of equilibrium of a pyranose or a furanose in its alpha and beta forms.

mutases
enzymes that catalyse the transposition of functional groups.

mutation
an inheritable change in the nucleotide sequence of a gene.

mutein
a mutant protein, especially a product of recombinant DNA technology.

myofibril
a unit of thick and thin filaments in a muscle fiber.

■ myokinase
a former name for adenylate kinase.

■ myosin
the main protein of the thick filaments in a muscle myofibril. It is composed of two coiled subunits (M_r about 220,000) that can aggregate to form a thick filament, which is globular at each end.

■ NAD
(nicotinamide adenine dinucleotide) Nicotinamide-containing coenzyme functioning as carriers of hydrogen atoms and electrons in some redox reactions, notably carbohydrate and lipid catabolism.

■ NADP
(nicotinamide adenine dinucleotide phosphate) Nicotinamide-containing coenzyme functioning as carriers of hydrogen atoms and electrons in some redox reactions, notably fatty acid synthesis and the pentose phosphate shunt.

■ naive
in immunology, descriptive of an unimmunised animal or a memory T-cell that has not been activated to respond to an antigen.

■ nascent
descriptive of a newly formed molecule during its formation or at the moment of its completion.

■ nascent RNA
the initial transcripts of RNA, before any modification or processing.

■ native conformation
the biologically active conformation of a protein.

■ native gel
a polyacrylamide gel for support of electrophoresis of proteins that is formulated without sodium dodecyl sulphate, so that the protein remains in its native form.

■ native structure
the structure of a protein or nucleic acid in which it is able to perform its physiological function.

■ negative control
repression of biological activity by the presence of a specific molecule.

negative cooperativity
a phenomenon of some multisubunit enzymes or proteins in which binding of a ligand or substrate to one subunit impairs binding of a second ligand molecule to another subunit.

negative feedback
regulation of a biochemical pathway achieved when a reaction product inhibits an earlier step in the pathway.

N-end rule
the observation that the half-life of a protein in vivo is a function of its N-terminal residue.

neo-
new.

Nernst equation
an equation that relates the redox potential to the standard redox potential and the concentrations of the oxidised and reduced form of the couple.

nerve gas
often a fluorophosphate inhibitor of acetylcholinesterase; chemically and functionally related to organophosphate insecticides.

nested deletion sequencing
a method for extension of the length of a target DNA that can be sequenced using a single primer. Removal by an appropriate nuclease of varying polydeoxynucleotide lengths from the region near the primer binding site on a plasmid brings into sequencing range different regions of the target, followed by religation and circularisation of the resultant shortened plasmids.

nested primers
a second set of PCR primers whose use follows an initial copying of a nucleic acid sequence; the nested primers are specific for sequences internal to the 3'- and 5'-ends of the originally copied sequence.

Neuberg ester
obsolete name for fructose 6-phosphate.

neuron
a cell of nervous tissue specialised for transmission of a nerve impulse.

neurotransmitter
a chemical signal that passes from an afferent nerve across the synapse to stimulate the efferent nerve, e.g. acetylcholine, noradrenaline (norepinephrine), dopamine.

neutraceutical
a dietary supplement.

ninhydrin reaction
a color reaction given by free amino groups of amino acids and peptides on heating with ninhydrin; widely used for their detection and estimation.

nitrogen cycle
the cycling of various forms of biologically available nitrogen (N_2, nitrate, nitrite, ammonia, etc.) through the plant, animal, and microbial worlds, and through the atmosphere and geosphere.

nitrogen fixation
the reduction of molecular nitrogen to ammonia and nitrogen-containing metabolites that is carried out synthetically, or naturally by blue-green algae or bacteria, e.g. Rhizobium species that inhabit nodules on the roots of legumes.

nitrogenase complex
a system of enzymes capable of anaerobic reduction of atmospheric nitrogen (N_2) to ammonia in the presence of ATP.

N-linked carbohydrate
a carbohydrate moiety of a glycoprotein that is attached by a glycosidic bond to the -amide nitrogen of an asparagine residue; may be complex-type or high-mannose-type.

noid
mind, spirit.

nomogram
the presentation of three parallel linear or non-linear scales of related values aligned so that a straight line between desired points on two of the scales also crosses the third scale at the corresponding value, e.g. a line through points on scales for centrifuge rotor diameter (cm) and speed (rev./min) also passes through the scale for centrifugal force (g) at the correct point.

■ **non-**
not.

■ **noncompetitive inhibition**
a type of enzyme inhibition not reversed by increasing the substrate concentration. The inhibitor binds to a site on the enzyme other than the active site.

■ **noncompetitive inhibitor**
an inhibitor of enzyme activity whose effect is not reversed by increasing the concentration of substratc molecule.

■ **noncyclic electron flow**
the light-induced flow of electrons from water to NADP in oxygen-evolving photosynthesis involving both photosystems I and II. Contrat with cyclic photophosphorylation.

■ **nonessential amino acids**
amino acids that can be made by humans and other vertebrates from simpler precursors, and thus not required in the diet.

■ **nonhaeme iron proteins**
proteins, most often acting in redox reactions, which contain iron but no porphyrin groups.

■ **nonpolar**
molecules or groups composed of elements not differing greatly in electronegativity that are poorly soluble in water.

■ **nonsense codon**
a codon that does not specify an amino acid, but instead signals the termination of a polypeptide chain (UAA, UAG, UGA).

■ **nonsense mutation**
a change in the base sequence that converts a sense codon (one that specifies an amino acid) to one that specifies a stop (a nonsense codon).

■ **nonsense suppressor**
a mutation, usually in the gene for a tRNA, that causes an amino acid to be inserted into a polypeptide in response to a termination codon.

■ **northern blotting**
by analogy with Southern blotting, a technique for de-

tection of specific RNA molecules; RNA from a cell is denatured and separated by slab gel electrophoresis, then blotted on to a sheet of nitrocellulose or nylon and hybridised with radiolabelled DNA that is complementary to the desired RNA, whose presence is subsequently indicated by autoradiography.

■ N-terminal

in a polypeptide sequence, that unique residue which is connected to the linear sequence by its carboxy group, leaving it with a free amino group. In practice, the amino group of an N-terminal residue may be modified, e.g. by acylation by an acetyl or fatty acyl group.

■ nuclear envelope

the structure that encloses the eukaryotic nucleus and consists of an inner and an outer nuclear membrane and the perinuclear space between the two.

■ nuclear receptor

a nuclear protein that allows regulation of gene expression by a lipophilic effector molecule, such as a steroid or thyroid hormone, eicosanoid, retinoid or ecdysone. When the effector molecule is bound to the receptor protein, the latter can also bind to a specific DNA sequence.

■ nuclease

an enzyme that cleaves phosphodiester bonds of nucleic acids.

■ nucleic acids

biologically occurring polynucleotides in which the nucleotide residues are linked in a specific sequence by phosphodiester bonds, either DNA and RNA.

■ nucleofuge

a chemical group that is displaced in a nucleophilic substitution or elimination reaction.

■ nucleohistone

a complex of DNA and histone.

■ nucleoid

in prokaryotes, the nuclear zone that contains the chromosome but has no surrounding membrane.

■ nucleolus
a spherical structure visible in the nucleus during interphase. The nucleolus is associated with a site on the chromosome that is involved in ribosomal RNA synthesis.

■ nucleophile
a compound or functional group that can attract and bind a proton or another positively charged species such as a carbocation ion.

■ nucleophilic group
an electron-rich group that tends to attack an electron-deficient nucleus.

■ nucleoplasm
the aqueous portion of a cell's contents enclosed by the nuclear membrane; also called the nuclear matrix.

■ nucleoprotein
a protein-DNA or protein-RNA complex.

■ nucleoside
a compound consisting of a purine or pyrimidine base covalently linked to a pentose, but lacking a phosphate.

■ nucleoside diphosphate kinase
an enzyme that catalyses the transfer of the terminal phosphate of a nucleoside 5'-triphosphate (most often ATP) to a nucleoside 5'-diphosphate (GDP, e.g.).

■ nucleoside diphosphate sugar
an activated form of a sugar molecule used in the enzymatic synthesis of polysaccharides and sugar derivatives. The base depends on the sugar: for example, UDP-glucose but GDP-fucose.

■ nucleoside monophosphate kinase
an enzyme that catalyses the transfer of the terminal phosphate of ATP to a nucleoside 5'-monophosphate.

■ nucleosome
structural unit for packaging DNA into chromatin. A nucleosome consists of a DNA strand wound around an octameric histone core.

■ nucleotide
an organic molecule containing a purine or pyrimidine base, a five-carbon sugar (ribose or deoxyribose), and one

or more phosphate groups. A phosphoester of a nucleoside.

■ **nucleotide-binding domain**
one of the homologous supersecondary structures of some proteins, composed of Rossmann folds, to which may be bound the AMP moiety of a substrate or ligand; e.g. the domain of a dehydrogenase that binds a pyridine nucleotide or a flavin-adenine dinucleotide.

■ **nucleus**
in eukaryotic cells, the centrally-located organelle that encloses most of the chromosomes. Minor amounts of chromosomal substance are found in some other organelles, most notably the mitochondria and the chloroplasts.

■ **nut site**
n utilisation site.

■ **O**
'Farrell gel.

■ **ob-**
before, against.

■ **octa-**
eight.

■ **octo-**
eight.

■ **oestrogen (estrogen)**
a compound, usually a steroid, that supports the development of female secondary sex characteristics; e.g. oestradiol.

■ **ogen**
precursor.

■ **Ogston hypothesis**
the proposal that an enzyme can distinguish in its substrate between two like substituents on a carbon atom that also carries two unlike groups, through binding sites for three of the four substituents, which forces the substrate to display in a unique configuration one of the like groups.

■ **oid**
resembling, image of.

■ **Okazaki fragment**
a short segment of single-stranded DNA that is an intermediate in DNA synthesis. In bacteria, Okazaki fragments are 1000-2000 bases in length; in eukaryotes, 100-200 bases in length.

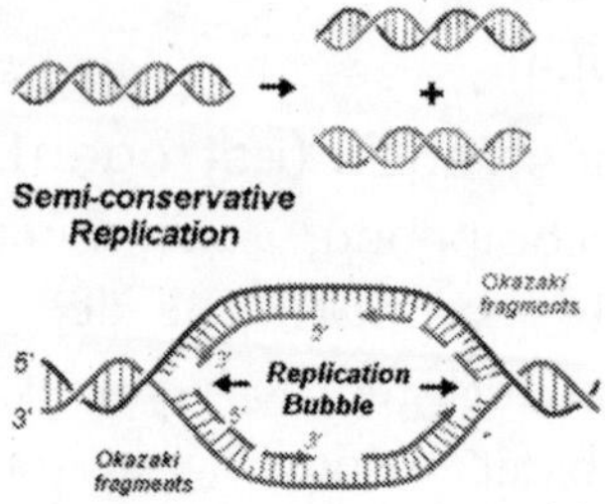

ol(e)
alcohol.

ole
little version(usually).

oligo-
few.

oligomer
a short polymer of amino acids, sugars, or nucleotides. The definition of "short" is somewhat arbitrary, but usually less than 50 nucleotides or amino acids or about a dozen monosaccharides.

oligomeric protein
a multisubunit protein having two or more polypeptide chains.

oligonucleotide
a polynucleotide containing a small number of nucleotides.The linkages are the same as in a polynucleotide; the only distinguishing feature is the small size.

oligonucleotide ligation assay (OLA)
a method to identify a specific oligonucleotide sequence without use of radiochemicals, electrophoresis or centrifugation. An oligonucleotide biotinylated at its 5'-end and another with a reporter group (chromophore or fluorophore) at its 3'-end are constructed to hybridise to the sequence to be detected in a template DNA strand. A ligase is able to join the two oligonucleotides if they both match the template perfectly. The ligated oligonucleotide can be extracted from the mixture on to immobilised streptavidin and the presence of the reporter group detected.

oligopeptide
a few amino acids (usually less than 50) joined by peptide bonds.

oligosaccharide
a molecule containing a small number of sugar residues

joined in a linear or a branched structure by glycosidic bonds.

■ oma

tumor.

■ omic research

following the currency of 'genomics', additional words have been coined to denote the compilation of data bases as resources for future research and analysis; proteonomics for proteins, kinomics for protein kinases, CHOnomics for carbohydrates, transcriptomics, vaccinomics, clonomics, functional, structural and pharmacogenomics, etc. Together, these approaches are termed omic research.

■ oncogene

a gene present in normal cells which when altered can transform the normal cell into a malignant cell. (When emphasizing this relationship, the unaltered gene is called a proto-oncogene and the altered gene, an oncogene.) The alteration may result in overproduction of a gene product or faulty function. An oncogene is denoted by the prefix c if a cellular oncogene, e.g. c-ras, or by the prefix v if a viral oncogene, e.g. v-ras, which is presumed to have originated from a capture of the c-oncogene by the virus during an infection of a normal cell. Contrasted with a thanatogene, or death gene, which when activated leads to apoptosis.

■ oncoprotein

the translation product of an oncogene; a regulator of cellular processes, which if the gene is mutated, results in uncontrolled growth (cancer). Oncoproteins may be embedded in the plasma membrane and serve as receptors, reside in the cytosol and be involved in signal transduction, often associated with a tyrosine kinase activity, or may be found in the nucleus and be directly involved in gene regulation.

■ one-carbon pool

the aggregate of freely interconvertible moieties that contain single carbons more reduced than carbon dioxide; i.e. metabolites of tetrahydrofolate, the methyl groups of choline, the hydroxymethyl of serine, etc.

■ **oocyte**

a female gamete, a precursor of an ovum.

■ **opal mutation**

a nonsense mutation that, due to the premature appearance in the mRNA of the terminator codon UGA, leads to the formation of a truncated, possibly non-functional, protein.

■ **open reading frame (ORF)**

a sequence of nucleotide codons in a DNA or RNA molecule that does not include a termination codon and thus could code for a protein.

■ **open system**

a system that exchanges matter and energy with its surroundings. See system.

■ **open-chain form**

the form of a sugar in which there has been no internal condensation to form a pyranose or furanose.

■ **operator**

a region of DNA that interacts with a regulatory protein to control the expression of a gene or group of genes.

■ **operon**

a group of contiguous genes that are coordinately regulated by two cis-acting elements, a promoter and an operator. Found only in prokaryotic cells.

■ **opsonin**

the immunoglobulin components of complement attached to a foreign cell that sensitise it to phagocytosis.

■ **optical activity**

the property of a molecule that leads to rotation of the plane of polarisation of plane-polarised light when the latter is transmitted through the substance. Chirality is a necessary and sufficient property for optical activity.

■ **optimum pH**

the characteristic pH at which an enzyme has maximal catalytic activity.

■ **optode**

a biosensor with a light-detecting sensor, e.g. a device that detects the lowering of pH due to the action of glucose oxidase at the end of a fibre optical light guide by the effect on a pH-sensitive fluorescent dye.

■ **organelle**
inclusion in the cell cytoplasm that can be membrane-limited (e.g. chloroplast, mitochondrion, endoplasmic reticulum, Golgi apparatus) or non-membrane-limited (e.g. cytoskeletal element, ribosome, nucleosome).

■ **origin**
the nucleotide sequence or site in DNA where DNA replication is initiated.

■ **orosomucoid**
an old term for $?_1$-acid glycoprotein.

■ **orphan receptor**
a membrane protein that has been discovered by cloning of its gene to be of the same family as known receptors, e.g. the G-protein-coupled receptors, but for which the activating ligand has not been identified. The process of identification of the natural ligand for orphan receptors has been called 'reverse physiology'.

■ **osazone**
the product of coupling the carbonyl carbon of a sugar with phenylhydrazine, its further oxidation and subsequent addition of a second phenylhydrazine to form a crystalline compound.

■ **osis**
full of.

■ **osmosis**
bulk flow of water through a semipermeable membrane into another aqueous compartment containing solute at a higher concentration.

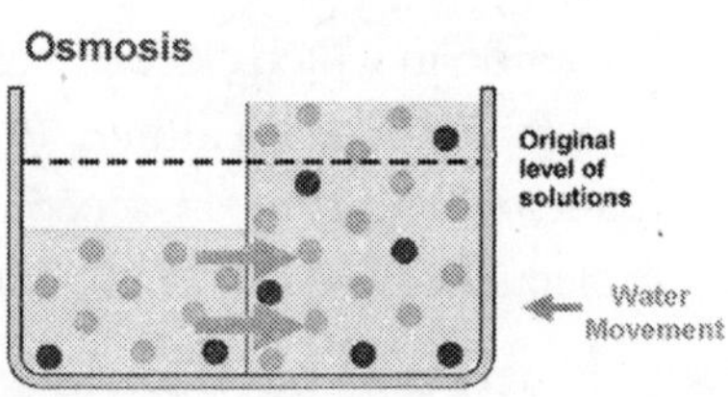

■ **osmotic pressure**
the pressure generated by the mass flow of water to that side of a membrane-bounded structure that contains the higher concentration of solute molecules. A stable osmotic pressure is seen in systems in which the membrane is not

permeable to some of the solute molecules.

overlapping gene

a rare phenomenon found in small viral genomes whereby a single mRNA can be translated in two different reading frames to produce two different proteins.

ovine

derived from sheep.

oxidation

the loss of electrons from a compound.

oxidation-reduction reaction (redox reaction)

a reaction in which electrons are transferred from a donor (the reducing agent) to an acceptor molecule (the oxidising agent).

oxidative decarboxylation

removal of carbon dioxide from a carboxylic acid that is facilitated by oxidation of the alpha -carbon (alpha -oxidative decarboxylation, e.g. the pyruvate dehydrogenase reaction) or the beta -carbon (beta -oxidative decarboxylation, e.g. the isocitrate dehydrogenase reaction).

oxidative phosphorylation

the reactions that synthesise the phosphoanhydride bond of ATP by the coupling of that energetically unfavourable reaction with the spontaneous oxidation of metabolites.

oxidative stress

the cumulative damage due to the less than 100% effectiveness of antioxidants in prevention of free radical reactions such as lipid peroxidation.

oxidising agent

the acceptor of electrons in an redox reaction.

oxidoreductase

one of a class of enzymes that oxidise one substrate as they reduce another, e.g. a dehydrogenase, an oxidase, a peroxidase.

oxygen debt

the extra oxygen (above the normal resting level) consumed in the recovery period after strenuous physical exertion due to the re-oxidation of lactate.

■ oxygen dissociation curve
a characterisation of an oxygen-binding protein, such as haemoglobin; a plot of percentage saturation against partial pressure of oxygen.

■ oxygenases
enzymes that catalyse reactions in which oxygen is introduced into an acceptor molecule.

■ p53
a tumor suppressor protein found to be mutated in many types of cancer.

■ packaging cell
in biotechnology, a eukaryotic cell that is used to produce a modified virus.

■ palindrome
a sequence of bases that reads the same in both directions on opposite strands of the DNA duplex (e.g., GAATTC).

■ paradigm
in biochemistry, an experimental model held to be true.

■ parenchyma
the functional tissue of an organ, as contrasted with the stroma.

■ parenteral nutrition
provision of food other than orally, usually by intravenous infusion.

■ partition chromatography
a technique for separation of molecules based on their differing ratios of solubility in two immiscible solvent phases. A stationary phase is coated or impregnated with one solvent phase and the mixture to be separated is passed over it in the mobile phase.

■ partition coefficient
a constant that expresses the ratio in which a given solute will be partitioned or distributed between two given immiscible liquids at equilibrium, or between a solvent and adsorbant.

■ passive diffusion
the transport of compounds across a membrane that is unmediated by any mechanism and is independent of energy sources, and therefore occurs at a rate that is determined by the area of the membrane, the concentration

difference across it and the solubility of the compound in the membrane.

■ **passive transport**
transport of a molecule across a membrane down a concentration gradient which does not require the input of energy.

■ **Pasteur effect**
the inhibition by oxygen of glucose consumption in a tissue or microbiological preparation.

■ **pathogenic**
disease-causing.

■ **pathy**
disease of, suffering.

■ **PCR (Polymerase Chain Reaction)**
a repetitive procedure that results in a geometric amplification of a specific DNA sequence. It involves using a thermostable DNA Polymerase such as Taq (Thermus aquaticus) and repeatedly heating to denature strands of DNA, followed by cooling to allow polymerisation of new strands.

■ **Pelham hypothesis**
the suggestion that aggregation of a globular protein occurs when the protein is unfolded, for instance by heat, and the specific intramolecular interactions that determined its native structure are replaced by the same interactions with other unfolded molecules of the same protein. In this view, heat-shock proteins help prevent this aggregation by competing with these intermolecular reactions and by somewhat weaker interactions, maintain the partially unfolded protein in an unaggregated state.

■ **pellet**
the sedimented portion that accumulates during centrifugation.

■ **penia**
lack.

■ **pent-**
five.

■ **pentose**
a simple sugar with a backbone containing five carbon atoms.

■ pentose nucleic acid
obsolete name for RNA.

■ pentose phosphate pathway
a pathway involved in the oxidation of glucose and is a source of reducing equivalents (NADPH) and pentoses for biosynthetic processes; present in most organisms. Also called the phosphogluconate pathway, pentose phosphate shunt, or hexose monophosphate shunt.

■ peptidase
an enzyme that hydrolyses a peptide bond.

■ peptide
an organic molecule in which a covalent amide bond is formed between the α-amino group of one amino acid and the α-carboxyl group of another amino acid, with the elimination of a water molecule. The resulting connection is called a peptide bond.

■ peptide bond
the amide bond formed by condensation of the α-carboxy group of one amino acid with the α-amino group of another; the bond that joins together the amino acid residues that comprise a protein, peptide or polypeptide.

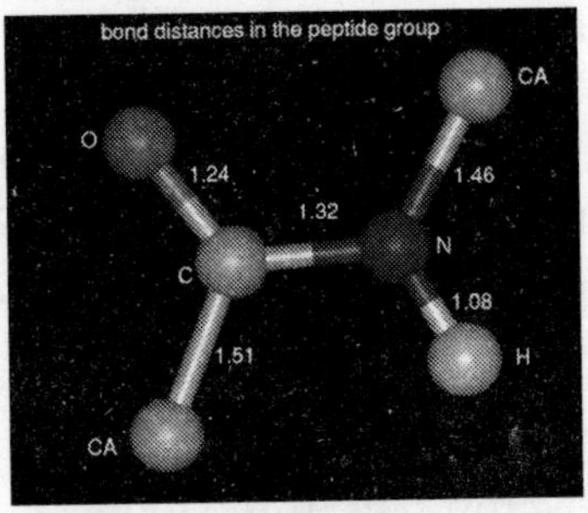

■ peptide mapping
the characteristic pattern of fragments formed by the separation of a mixture of peptides resulting from hydrolysis of a protein or peptide. AKA peptide finger-printing.

■ peptidoglycan
the main component of the bacterial cell wall, consisting of a two-dimensional network of heteropolysaccharides running in one direction, cross-linked with polypeptides running in the perpendicular direction.

■ peptoid
an alalog of a polypeptide, a poly-glycine in which the backbone nitrogen atoms are substituted with groups the same as or different from

those of natural amino acids. Although these polymers do not have chiral centres, their N-substituted amide bonds are set into cis or trans conformations, which allows them to adopt helical structures, somewhat like –helices.

■ per-
by, through, throughout.

■ perfusion chromatography
a variant of high-pressure liquid chromatography in which particles as the stationary phase have pores large enough [6000-8000Å (600-800nm)] to permit solvent flow-through and thus allow much faster flow rates, decreased diffusional spreading and enhanced access to the full surface of the stationary phase.

■ peri-
around, round-about.

■ peripheral nervous system (PNS)
a division of the nervous system of higher animals that consists of the neurons that extend from the spinal cord to skeletal muscles, glands, etc. and the neurons that return from these structures to the spinal cord.

■ peripheral proteins
proteins that are loosely or reversibly bound to a membrane by hydrogen bonds or electrostatic forces which can be easily released from the membrane. Contrast with integral proteins.

■ periplasm
the region between the inner (cytoplasmic) membrane and the cell wall or outer membrane of a bacterium.

■ permeable
the property of allowing material to pass through, as a permeable membrane.

■ permease
a protein that catalyses the transport of a specific small molecule across a membrane.

■ peroxidase
a haem enzyme that abstracts two hydrogens from one substrate to reduce hydrogen peroxide, its other substrate, to water.

■ peroxisome

membrane-bounded organelle in the cytoplasm of eukaryotic cells functioning in the destruction of proteins and characterised by peroxide-forming enzymes.

■ pexy

fix in place.

■ pH

a measure of acidity of aqueous solutions; the negative of the common logarithm of the hydrogen ion concentration, -log[H+]. Values below 7 are acidic; values above 7 are alkaline.

■ phage

bacteriophage.

■ phage display

a method for incorporation, into the DNA that encodes a phage coat protein, of a polypeptide, where it is available for a selection procedure; especially for production and selection of antibody (even anti-self antibody) fragments directed against a specific antigen. A library of VH or VL genes is produced by PCR using primers directed to constant regions that flank the variable region, cloned, and inserted into the gene for a phage coat protein that, when expressed, is available for cycles of growth and selection by adsorption on to an antigen-coated plate or beads.

■ phagosome

a vesicle that forms in a cell as a result of phagocytosis and that contains the engulfed material.

■ phase problem

in X-ray crystallography, the gap between the requirements for reconstruction of a model - for each reflection (defined by the angle of its diffraction), the amplitude (related to intensity) and phase (a measure of the alignment of the wave function with that of other reflections) - and the experimentally accessible quantities, the intensities of each reflection. (see also diffraction pattern; Fourier transformation).

■ phenotype

the observable trait(s) that result from the genotype in cooperation with the environment.

phenylketonuria
a human disease caused by a genetic deficiency in the enzyme that converts phenylalanine to tyrosine. The immediate cause of the disease is an excess of phenylalanine, which can be alleviated by a diet low in phenylalanine.

pheromone
a chemical signal that travels through air or water from one organism to another.

-philia
affection for.

phosphagen
a compound that acts as a storage form of high-energy phosphate, e.g. phosphocreatine, phosphoarginine.

phosphodiester
a molecule containing two alcohols esterified to a single molecule of phosphate. For example, the backbone of nucleic acids is connected by 5'-3' phosphodiester linkages between the adjacent individual nucleotide residues.

phosphodiester linkage
a chemical grouping that contains two alcohols esterified to one molecule of phosphoric acid, which thus serves as a bridge between them. Example

phosphogluconate pathway
an oxidative pathway beginning with glucose-6-phosphate and leading to ribose and NADPH. Also called the pentose phosphate pathway or hexose monophosphate shunt.

phospholipid
a lipid containing charged hydrophilic phosphate groups; a component of cell membranes.

phosphorolysis
cleavage of a compound with phosphate as the attacking group: analogous to hydrolysis.

phosphorylation
formation of a phosphate derivative of a biomolecule, most often by enzymatic transfer of a phosphate group from ATP.

phosphorylation potential
the actual free-energy change of ATP (or other phosphate-

containing molecule) hydrolysis under the nonstandard conditions within a cell.

photochemical reaction center
the part of a photosynthetic complex where the energy of an absorbed photon causes charge separation, initiating electron transfer. The "special pair"

photon
a unit of light energy.

photophosphorylation
the enzymatic formation of ATP from ADP coupled to the light-dependent transfer of electrons from water in photosynthetic cells.

photoreactivation
dNA repair in which the damaged region is repaired with the help of light and an enzyme. The lesion (break) is repaired without excision (cutting out) from the DNA.

photoreduction
the light-induced reduction of NADP in photosynthetic cells.

photosynthesis
the biosynthesis that directly harnesses the chemical energy resulting from the absorption of light. Frequently used to refer to the formation of carbohydrates from CO_2 that occurs in the chloroplasts of plants or the plastids of photosynthetic microorganisms.

photosystem
in photosynthetic cells, a functional set of light-absorbing pigments and its reaction center (antenna molecules, a special pair and the associated proteins).

phototroph
an organism that can use the energy of light to synthesise its own fuels from simple molecules such as carbon dioxide, oxygen, and water. Contrast with chemotroph.

Pitch length (or pitch)
the number of base pairs per turn of a duplex helix.

pK
the negative logarithm of an equilibrium constant.

■ **plaque**
a circular clearing on a lawn (continuous layer) of bacterial or culture cells, resulting from cell Iysis and production of phage or animal virus progeny.

■ **plasma membrane**
the exterior membrane surrounding the cytoplasm of a cell.

■ **plasma proteins**
the proteins present in blood plasma: Albumin, antibodies, etc.

■ **plasmalogen**
a phospholipid with an alkenyl ether substituent on the C-1 of glycerol. [A plasmalogen]

■ **plasmid**
a circular DNA duplex that replicates autonomously in bacteria. Plasmids that integrate into the host genome are called episomes. Plasmids differ from viruses in that they never form infectious nucleoprotein particles.

■ **plastid**
in plants, a self-replicating organelle; may differentiate into a chloroplast.

■ **plasty**
re-shaping.

■ **platelets**
small, enucleated cells that initiate blood clotting which arise from megakaryocytes in the bone marrow. AKA thrombocytes.

■ **pleated sheet**
the side-by-side, hydrogen-bonded arrangement of polypeptide chains in the extended ß conformation.

■ **pleo-**
more than usual.

■ **polar**
molecules which contain elements differing in electronegativity, making them soluble in water.

■ **polar group**
a hydrophilic (water-loving) group.

■ **polar mutation**
a mutation in one gene that reduces the expression of a gene or genes distal to the promoter in the same operon.

■ **polarimeter**
an instrument for determining the rotation of polarisation of

light as the light passes through a solution containing an optically-active substance.

polarity

depending on the context meaning one of two things. Either the nonuniform distribution of electrons in a molecule resulting from elements with differing electronegativities bonding; or the distinction between the 5' and 3, ends of nucleic acids.

poly

many.

polyamine

a hydrocarbon containing more than two amino groups.

polycistronic messenger rna

in prokaryotes, an RNA that contains two or more cistrons; note that only in prokaryotic mRNAs can more than one cistron be utilised by the translation system to generate individual proteins.

polyclonal antibodies

a heterogeneous pool of antibodies produced in an animal by a number of different B lymphocytes in response to an antigen. Different antibodies in the pool recognise different epitopes of the antigen.

polylinker

a short, usually synthetic, fragment of DNA containing recognition sequences for several restriction endonucleases employed to insert DNA into plasmids.

polymerase

an enzyme that catalyses the synthesis of a polymer from monomers.

polymerase chain reaction (PCR)

a repetitive procedure that results in a geometric amplification of a specific DNA sequence. It involves using a thermostable DNA Polymerase such as Taq (Thermus aquaticus) and repeatedly heating to denature strands of DNA and cooling to allow polymerisation of new strands.

polymorphic

describing a protein for which amino acid sequence variants exist in a population of organisms, but the variations do not destroy the protein's function.

For example, antibodies which contain variable and hypervariable sites.

- **polynucleotide**

  a chain structure containing nucleotides linked together by phosphodiester (5'-3') bonds. The polynucleotide chain has a directional sense with a 5' and a 3' end.

- **polynucleotide phosphorylase**

  an enzyme that polymerises ribonucleotide diphosphates.

- **polypeptide**

  a long chain of amino acids (less than 100) linked by peptide bonds.

- **polyribosome (polysome)**

  a complex of an mRNA and two or more ribosomes actively engaged in protein synthesis.

- **polysaccharide**

  a linear or branched polymer of monosaccharide units linked by glycosidic bonds.

- **polysome (polyribosome)**

  a complex of an mRNA molecule and many ribosomes.

- **porphyrin**

  a complex planar structure containing four substituted pyrroles covalently joined in a ring and frequently containing a central metal atom. For example, haeme is a porphyrin with a central iron atom.

- **positive control**

  a system that is turned on by the presence of a regulatory protein.

- **positive cooperativity**

  a phenomenon of some multisubunit enzymes or proteins in which binding of a ligand or substrate to one subunit facilitates binding of a second ligand or substrate to another subunit.

- **post-**

  behind, after.

- **posttranscriptional processing**

  the enzymatic processing of an hnRNA, producing functional mRNA, tRNA, or rRNA molecules.

- **posttranslational modification**

  enzymatic processing of a polypeptide chain after translation from its mRNA. Examples are enzymatic cleavage, phosphorylation or sulphation, and oligosaccharide addition.

■ pre-
before, in front.

■ primary structure
the sequence of amino acids and any interchain and intrachain disulphide bonds of a protein.

■ primary transcript
the immediate RNA product of transcription before any posttranscriptional processing reactions. Heterogenous nuclear (hnRNA) is also used.

■ primase
an enzyme that synthesises RNA oligonucleotides used as primers by DNA polymerases.

■ primer
a structure that serves as a growing point for polymerisation.Short primers of DNA are often used in sequencing and mutagenesis procedures.

■ primosome
a multiprotein complex that catalyses synthesis of RNA primer at various points along the DNA template.

■ probe
a labeled fragment of nucleic acid containing a nucleotide sequence complementary to a nucleotide sequence that one wishes to detect in a hybridisation experiment.

■ processivity
for any enzyme that catalyses the synthesis of a biological polymer, the property of adding multiple subunits to the polymer without dissociating from the substrate.

■ prochiral molecule
a nonchiral molecule that lacks handedness and is optically inactive, but would become chiral by a change in one of the substituents at the chiral center. A prochiral molecule may react with an enzyme so that two groups that have a mirror-image relationship to each other are treated differently.

■ prokaryote
a unicellular organism that contains a single chromosome, no nucleus, no membrane-bound organelles, and

has characteristic ribosomes and biochemistry.

■ **promoter**
that region of the gene that signals RNA polymerase binding and the initiation of transcription.

■ **prophage**
the silent phage genome. Some prophages integrate into the host genome; others replicate autonomously.The prophage state is maintained by a phage-encoded repressor.

■ **prophase**
the stage in meiosis or mitosis when chromosomes condense and become visible as refractile bodies.

■ **proprotein**
a protein that is made in an active form, so that it requires processing to become functional.

■ **prostaglandin**
an oxygenated eicosanoid that has a hormonal function. Prostaglandins are unusual hormones in that they usually have effects only in that region of the organism where they are synthesised.

■ **prosthetic group**
a metal ion or an organic compound (other than an amino acid) that is covalently bound to a protein and is essential to its activity.

■ **protamines**
highly basic, arginine-rich proteins found complexed to DNA in the sperm of many invertebrates and fish.

■ **protein**
a macromolecule composed of one or more polypeptide chains. Each polypeptide chains has a characteristic sequence of amino acids linked by peptide bonds. The molar mass is usually above 100,000

■ **protein kinases**
enzymes that phosphorylate certain amino acid residues (most often Ser, Thr, or Tyr) in specific proteins.

■ **protein subunit**
one of the components or monomers of a multicomponent protein.

■ **protein targeting**
the process by which newly synthesised proteins are sorted and transported to

their proper locations in the cell.

■ proteoglycan

1. a hybrid macromolecule consisting of a larger heteropolysaccharide joined to a smaller polypeptide, i.e. the polysaccharide is the major component.

2. a protein-linked heteropolysaccharide in which the heteropolysaccharide is usually the major component.

■ protist

a relatively undifferentiated organism that can survive as a single cell.

■ proton acceptor

a functional group capable of accepting a proton from a proton donor molecule.

■ proton donor

the donor of a proton in an acid-base reaction: a Bronsted-Lowry acid.

■ proton-motive force

the electrochemical potential inherent in a transmembrane H^+ concentration gradient. The proton gradient is used in oxidative phosphorylation and photophosphorylation to drive ATP synthesis.

■ proto-oncogene

a gene of cellular or viral origin, usually encoding a regulatory protein, that can be converted into an oncogene by mutation.

■ prox-

besides.

■ pseudo-

false.

■ pseudocycle

a sequence of reactions that can be arranged in a cycle but that usually do not function simultaneously in both directions. Also called a futile cycle, since the net result of simultaneous functioning in both directions would be the expenditure of energy without accomplishing any useful work.

■ pulse-chase

an experiment in which a short labeling period is followed by the addition of an excess of the same, unlabeled compound to dilute out the labeled material. Useful for observing time-dependent behavior of compounds.

■ purine

a nitrogenous heterocyclic base found in nucleotides and nucleic acids; containing fused pyrimidine and imidazole rings, having the general form.

■ puromycin

an antibiotic that inhibits polypeptide synthesis by being incorporated into a growing polypeptide chain, causing its premature termination.

■ pyranose

a simple sugar containing the six-membered pyran ring.

■ pyridine nucleotide

a nucleotide coenzyme containing the pyridine derivative nicotinamide: NAD(H) or NADP(H).

■ pyridoxal phosphate

a coenzyme containing the vitamin pyridoxine (vitamin B_6) whose function is in reactions involving amino group transfer.

■ pyrimidine

a heterocyclic six-membered ring structure. Cytosine and uracil are the main pyrimidines found in RNA, and cytosine and thymine are the main pyrimidines found in DNA.

■ pyrimidine dimer

a covalently joined dimer of two adjacent pyrimidine residues in DNA, induced by absorption of UV light, usually derived from two adjacent thymines (a thymine dimer).

■ pyrophosphate

a molecule formed by two phosphates in anhydride linkage.

■ quadrapole mass filter

the instrumentation for mass spectroscopy that allows selection of ions of a specific m/e- ratio. As the desorbed material travels down an electric field toward the detector, it passes through a space boarded by four parallel rods, which are electrodes. All four electrodes have the same DC current, onto which is superimposed an AC current that oscillates at a rate calculated to retain within the four rods only ions of the desired m/e- ratio and to deflect all other ions. Thus, the signal-to-noise ratio is kept high. By varying the charges on the electrodes, the spectrum of ions can be sampled. An alternative design is the ion-trap, in which the desorbed ions enter a cavity bounded front and back by circular entrance and exit endcap electrodes, and on the sides by a ring electrode. The voltages on all three electrodes are varied so that incoming ions, passing through aperture in the entrance endcap electrode are directed into a circular orbit inside the ring electrode, until a sufficient number are accumulated, at which time, they are directed through an aperture in the exit endcap electrode toward the detector in a sequence that reflects their m/e- ratio.

■ quadruplex DNA

a form of DNA in which four oligo(G) sequences, either of the same or of different strands, line up either in a parallel, an anti-parallel or in a fold-back (mixed parallel and antiparallel) pattern. The interior of quadruple the helix has four guanine bases in Hoogstein base pairings; each plane of four guanines is separated from the adjacent plane by a Na^+ or K^+ ion. Although these structures have not been identified cells, they are strongly suspected to cap the ends of chromosomal DNA.

■ quantum yield

the ratio of the number of molecules that respond, e.g. by reacting or fluorescing, to the number of photons absorbed.

■ quaternary structure

in a protein, the way in which the different folded subunits interact to form the multisubunit protein.

■ quiescent

non-proliferating, as applied to cells.

■ quiescent affinity label

a reagent that reacts specifically with the active site of the enzyme for which it was designed. It bears a generally reactive chemical group, such as a chloromethane, and substrate-like moieties that permit it to bind only to the target enzyme.

quinimine form
a tautomer of the Schiff base adduct of pyridoxal phosphate with an amino acid in which the pyridine ring nitrogen is protonated but uncharged and the carbon atom of the coenzyme that attaches the amino group of the amino acid has lost a proton. It is an intermediate between the aldimine form (in which the pyridine ring is protonated and charged and the attachment of the amino acid to the cofactor is a Schiff base) and the ketimine form (which can best be represented as a Schiff base formed by condensation of an a-oxo acid with pyridoxamine).

quinoprotein
a protein, usually of microbial origin, with a quinone prosthetic group: pyrroloquinoline quinone (e.g. methanol dehydrogenase), 6-hydrodopaquinone (topa quinone, e.g. amine oxidase) or 2',4-bitryptophane-6, 7-dione (tryptophan tryptophanylquinone, e.g. amine dehydrogenase).

quorum-sensing
the ability of some micro-organisms to monitor cell density cues, such as acylhomoserine lactones, and to adjust their gene expression accordingly.

R group
an abbreviation used to denote an organic substituent such as the R groups of amino acids.

R loop
a triple-stranded structure in which RNA displaces a DNA strand by DNA-RNA hybrid formation in a region of the DNA.

racemic mixture
a mixture of the D and L stereoisomers of an optically active compound.

radical
an atom or group of atoms possessing an unpaired electron.

radioactive isotope
an isotopic form of an element with an unstable nucleus that stabilises itself by emitting ionising radiation

and can be detected for use as a tracer.

■ radioimmunoassay
a sensitive and quantitative method for detecting trace amounts of a biomolecule, based on its capacity to displace a radioactive form of the molecule from combination with its specific antibody.

■ Ramachandran plot
a graph that displays calculated energy isotherms for conformations of peptide backbones; phi-angles against psi-angles; energy minima correspond to beta-structure, alpha-helix and the collagen triple helix.

■ rapid-start complex
the complex that RNA polymerase forms at the promoter site just before initiation.

■ ras protein
a product of the ras gene; a G-protein.

■ rate-limiting step
the slowest step in a metabolic pathway.

■ re, red-
back, again.

■ reaction centre
in photosynthesis, the energy-transducing unit. By absorption of light it excites a chlorophyll dimer, which can then transfer a pair of electrons across a membrane to a quinone, which will accept two protons from its microenvironment to generate a hydroquinone.

■ reaction co-ordinate
a graphical representation of the course of a reaction that shows the energy of the reactants and products as the reaction progresses.

■ reactive oxygen
the group of chemically highly reactive species that are related to molecular oxygen, i.e. O_2-• (superoxide), HO• (hydroxyl radical) and H_2O_2 (hydrogen peroxide).

■ reading frame
a sequence of three-nucleotide codons in DNA or RNA.

■ receptor
the binding site for a hormone or neurotransmitter that initiates its action at the cellular level.

■ recombinant DNA
DNA formed by bringing together DNA fragments from different species.

■ recombination
the natural or synthetic production of a DNA molecule from polynucleotides derived from more than one parent DNA molecule.

■ red drop
the sharp decline in photosynthetic efficiency as the wavelength of exciting light enters the red region (above 680nm) due to the insensitivity of photosystem II to excitation by red light.

■ redox
descriptive of a chemical or biochemical reaction that involves oxidation of one reactant or substrate at the expense of reduction of another.

■ redox couple
an electron donor and its corresponding oxidised form.

■ redox pair
an electron donor and its corresponding oxidised form; for example, NADH and NAD^+.

■ redox potential (E)
the relative tendency of a pair of molecules to release or accept an electron. The standard redox potential (E^0) is the redox potential of a solution containing the oxidant and reductant of the couple at standard concentrations.

■ redox reaction
oxidation-reduction reaction.

■ reducing agent (reductant)
the electron donor in a redox reaction.

■ reducing end
the end of a polysaccharide having a terminal sugar with a free anomeric carbon; the terminal residue can act as a reducing sugar during the portion of time it exists in the open-chain form.

■ reducing equivalent
a general term for an electron or an electron equivalent, usually in the form of a hydrogen atom or a hydride ion.

■ reducing sugar
a sugar in which the anomeric carbon is not involved in a glycosidic bond and can there-

fore undergo oxidation during the portion of time it exists in the open-chain form.

■ reduction
the gain of electrons.

■ regulatory enzyme
an enzyme in which the active site is subject to regulation by factors other than the enzyme substrate. The enzyme frequently contains a nonoverlapping site for binding the regulatory factor that affects the activity of the active site.

■ regulatory gene
a gene that gives rise to a product involved in the regulation of the expression of another gene. An example would be the gene coding for cro, a repressor protein.

■ regulatory sequence
a DNA sequence involved in regulating the expression of a gene: a promoter or operator.

■ regulon
a group of genes that are coordinately regulated.

■ reiteration
repetition of identical, or nearly identical, long DNA sequences in a chromosome.

■ relaxation
in kinetics and spectroscopy, the return to equilibrium of a macromolecule after a very brief perturbation of the environment, e.g. recovery of a protein from a sudden burst of energy that allows a fast transient increase in temperature (temperature-jump) or a fast transient change of pH (pH-jump).

■ release factors
1. termination factors.
2. hypothalamic hormones that stimulate release of other hormones by the pituitary gland. For example, Luteinising Hormone Releasing Hormone (LHRH).

■ renaturation
the process of returning a denatured structure to its original native structure, as when two single strands of DNA are reunited to form a regular duplex, or an unfolded polypeptide chain is returned to its normal folded three-dimensional structure.

■ renaturation
the return to native structure from a denatured state; in

nucleic acid chemistry, identical to annealing.

rennin

obsolete name for chymosin, a milk-clotting aspartate proteinase secreted by the stomachs of newborn mammals; not to be confused with renin.]

repair synthesis

DNA synthesis following excision (cutting out) of damaged DNA.

repertoire cloning

a method to produce human monoclonal antibodies. mRNAs are isolated from a tissue that generates immunoglobulins (e.g. spleen) and a reverse transcriptase makes cDNA copies of them. The Fd regions, i.e. the H-chain component of the Fab fragment (the VH and CH1 domains of the immunoglobulins regions), are selected and amplified by PCR to obtain a heavy-chain library. Similarly, a light-chain library is obtained. The two populations of cDNA are combined into a random pairing to form combinatorial constructs, inserted into coliphage vector along with a promoter, and used to infect E. coli. Plaques are then screened for ability to bind an antigen.

repetitive DNA

sequences of DNA that are found to be repeated, sometimes thousands of times over.

replicaplating

a technique in which an impression of a culture is taken from a master plate and transferred to a fresh plate. The impression can be of bacterial clones or phage plaques.

replication

synthesis of two daughter DNA molecules identical to the parental DNA.

replication fork

the Y-shaped region of DNA at the site of DNA synthesis; also called a growth fork.

replicon

a genetic element that behaves as an autonomous replicating unit. It can be a plasmid, phage, or bacterial chromosome.

replisome

the multiprotein complex that promotes DNA synthesis at the replication fork.

■ **repressible enzyme**
an enzyme whose synthesis is inhibited when its reaction product is readily available to the cell. Occurs only in bacteria.

■ **repression**
a decrease in the expression of a gene in response to a regulatory protein.

■ **repressor**
a regulatory protein that inhibits transcription from one or more genes. It can combine with an inducer (resulting in specific enzyme induction) or with an operator element (resulting in repression).

■ **residue**
a single unit within a polymer, for example, an amino acid within a polypeptide chain.

■ **resolution**
in X-ray crystallography, the precision with which atoms are located in space; usually expressed in Å (10^{-10}m).

■ **resonance energy transfer**
a phenomenon in which one fluorophore excites another when there is significant overlap of the emission spectrum of the primary fluorophore with the excitation spectrum of the secondary fluorophore and some other conditions are met. In part it depends strongly upon the distance that separates the fluorophores and their orientation towards one another; it is useful in calculation of intermolecular distances of macromolecules.

■ **resonance hybrid**
a molecular structure that is a hybrid of two structures that differ in the locations of some of the electrons. For example, the benzene ring can be drawn in two ways, with double bonds in different positions. The actual structure of benzene is in-between these two equivalent structures.

■ **respiration**
the catabolic process in which electrons are removed from nutrient molecules such as glucose and passed through a chain of carriers to oxygen.

■ **respiration state**
the condition of tightly coupled mitochondria that is

characterised by their respiratory rate (all but state 3 are slow) and defined by the rate-limiting component: state 1, ADP- and substrate-limiting; state 2, substrate-limiting; state 3, respiratory chain-limiting; state 4, ADP-limiting; state 5, oxygen-limiting.

■ **respiratory chain**

the sequence of electron-carrying proteins that transfer electrons from substrates to molecular oxygen in aerobic cells.

■ **resting potential**

the membrane potential of an excitable cell in the absence of stimulation.

■ **restriction endonucleases**

site-specific endonucleases causing cleavage of both strands of DNA at points within or near the specific site recognised by the enzyme. Many generate sticky ends, making them important tools in genetic engineering.

■ **restriction fragment**

a segment of double-stranded DNA produced by the action of a restriction endonuclease on a larger DNA.

■ **restriction fragment length polymorphism**

variations, among individuals in a population, in the length of certain restriction fragments within which certain genomic sequences occur. These variations result from rare sequence changes that create or destroy restriction sites in the genome.

■ **restriction-modification system**

a pair of enzymes found in most bacteria (but not eukaryotic cells). The restriction enzyme recognises a certain sequence in duplex DNA and makes one cut in each unmodified DNA strand at or near the recognition sequence. The modification enzyme methylates (or modifies) the same sequence, thus protecting it from the action of the restriction enzyme.

■ **retentate**

non-filtered particles and macromolecules.

■ reticulocyte
a non-nucleated (in mammals) precursor of an erythrocyte, still capable of haemoglobin synthesis.

■ retinol-binding protein
21 kDa protein.

■ retro-
backwards, behind.

■ retrovirus
an RNA virus containing reverse transcriptase.

■ reverse transcriptase
an RNA-directed DNA polymerase in retroviruses. These viruses are capable of making DNA complementary to an RNA and incorporating it into the genome.

■ rhage
burst out.

■ rhea
discharge, flowing out.

■ rhexis
shredding.

■ rho factor
a protein involved in the termination of transcription of some messenger RNAs.

■ ribonuclease
a nuclease that catalyses the hydrolysis of specific phosphodiester linkages of RNA.

■ ribonucleic acid
RNA.

■ ribonucleotide
a nucleotide containing D-ribose as its pentose component.

■ ribophorin
a protein of the rough endoplasmic reticulum that recognises leader sequences of growing polypeptides and, through them, attaches the ribosomes.

■ ribose
the five-carbon sugar found in RNA.

■ ribosomal RNA (rRNA)
a class of RNA molecules serving as major structural components of ribosomes.

■ ribosome
a supramolecular complex of rRNAs and proteins. Composed of a large and a small subunit which together are the site of protein synthesis.

ribosome jumping
5'-Cap-dependent initiation of translation of some viral mRNAs in which a ribosome subunit recognises the 5'-cap of the mRNA but 'jumps' over the 5'-non-translated region of strong secondary structure to reach the AUG start codon.

ribosomes
small cellular particles made up of ribosomal RNA and protein. They are the site, together with mRNA, of protein synthesis.

ribozymes
ribonucleic acid molecules with catalytic activities.

right-handed helix
by analogy with the right hand, a helix, e.g. of a polypeptide, that advances towards the C-terminus in the direction of the extended thumb as the backbone of a chain turns in the direction of the fingers closing on the palm. A left-handed helix has the opposite sense.

RNA chaperone
a putative RNA-binding protein that assists RNA in attainment of its biologically active conformation. Non-specific RNA-binding proteins are proposed to prevent RNA becoming temporarily trapped in innumerable non-functional metastable conformations (kinetic trapping); specific RNA-binding proteins are proposed to stabilise the biologically active conformation.

RNA polymerase
an enzyme that catalyses the formation of RNA from NTPs, using a strand of DNA or RNA as a template.

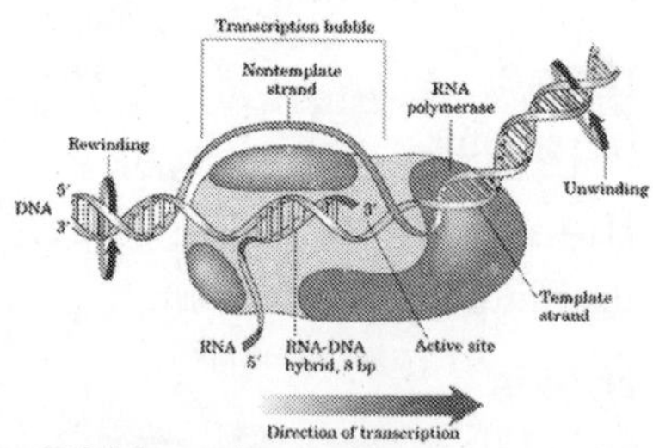

(a)

RNA splicing
removal of introns and joining of exons in a primary transcript (hnRNA).

RNP
RNA-protein complex.

■ ROA
Raman optical activity.

■ Robison ester
obsolete name for glucose 6-phosphate.

■ rolling circle replication
mechanisms of DNA replication of small circular genomes, such as some viruses and plasmids, in which the template generates multiple tandem copies. One strand is nicked and serves as a primer for elongation at the 3'-end, while the unnicked strand serves as a template.

■ rolling transcription
the synthesis of RNA from a circular DNA template. Some circular DNAs, even without a promoter sequence and without assistance of a primer, may be transcribed to form concatenated circular RNA. If the circular DNA encodes a ribozyme, the concatenates may process themselves to form monomeric RNA.

■ root effect
a phenomenon in some teleost fish that allows them to generate O_2 gas in their swim bladders in the face of considerable hydrostatic pressure. The haemoglobin structure and the local secretion of lactic acid create an exaggerated Bohr effect.

■ Rossmann fold
a supersecondary structure found in some proteins; a motif of beta strands connected by alpha-helix crossover elements.

■ rRNA
a class of RNA molecules serving as components of ribosomes.

■ run-on assay
a method for evaluation of the status of replication, transcription or translation in a cell. The assay continues the synthesis of a polymer without reinitiation.

■ S phase
the period during the cell cycle when the chromosome is replicated.

■ saccharide
a sugar or a polymer of sugars linked by glycosidic (acetal or ketal) bonds.

■ S-adenosylmethionine

(adoMet or SAM)
an enzymatic cofactor involved in methyl group transfers.

■ **salting in**
the increase in solubility that is displayed by typical globular proteins upon the addition of small amounts of certain salts, such as ammonium sulphate.

■ **salting out**
the decrease in protein solubility that occurs when salts such as ammonium sulphate are present at high concentrations.

■ **salvage pathway**
a family of reactions that permits, for instance, nucleosides as well as purine and pyrimidine bases resulting from the partial breakdown of nucleic acids to be re-utilised in nucleic acid synthesis.

■ **saponification**
alkaline hydrolysis of triacylglycerols to yield fatty acid salts and glycerol.

■ **saponification number**
a characteristic of an acylglycerol; the number of mg of KOH necessary to neutralise the fatty acids liberated from 1g of material.

■ **sarcolemma**
the excitable plasma membrane of a muscle cell.

■ **sarcomere**
the minimal functional unit of skeletal muscle; roughly a cylinder bounded on its ends by Z-lines where the thin filaments are anchored; these extend towards the centre. Thick filaments extend from the centre of the sarcomere, forming the A-region. The thick and thin filaments overlap in two darker-staining zones, which bracket the central H-region.

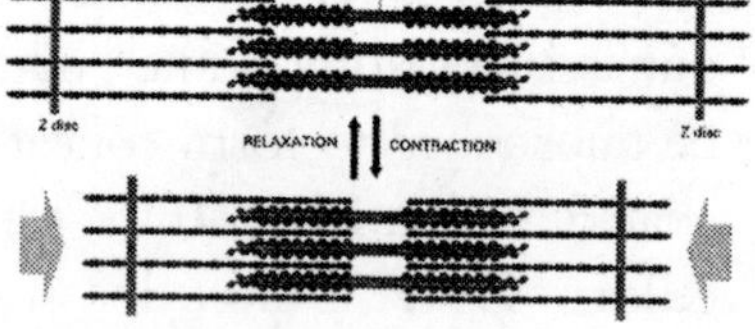

■ **sarcosome**
a mitochondrion of a muscle cell.

■ **satellite DNA**
a DNA fraction whose base composition differs from that

of the main component of DNA, as revealed by the fact that it bands at a different density in a CsCI gradient. Usually repetitive DNA or organelle DNA.

■ saturated fatty acid

a fatty acid containing no double bonds.

■ scaffold protein

one of the non-enzymic components of a multi- protein signal transduction complex. The complex is assembled upon activation of the receptor or is assembled but dormant until receptor activation. The scaffold is the framework for assembly at the cytoplasmic domain of a receptor; with the assistance of anchoring proteins it recruits kinases, phosphatases and other enzymes, and, with the assistance of adaptor proteins, other factors which will continue the signal sequence within the cell.

■ Scatchard plot

a graphical method for determination of a binding constant and the number of binding sites. The ratio of the amount of ligand (e.g. a hormone) that is bound (e.g. to a receptor) to the amount that is free is plotted against the free amount.

■ Schardinger dextrin

a product of bacterial degradation of starch; a closed-ring oligosaccharide in which glucosyl monomers are joined by a1-à4 glycosidic bonds.

■ Schiff base

the product formed by the reversible condensation of an aldehyde with an amino group in which the elements of water have been eliminated. It is featured in structures of biochemical interest, notably the reactions of amino acids with pyridoxal phosphate as a coenzyme of various enzymes of amino acid metabolism.

■ Schlieren method

an optical method for detection of the non-uniform dispersion of a macromolecule in solution, especially during analytical ultracentrifugation or free-zone electrophoresis, by the refractive index of its solution.

■ **scissile bond**
the bond of a substrate that is subject to enzymic cleavage.

■ **second law of thermodynamics**
the relationship of free energy (G), enthalpy (H) and entropy (S) when a system changes from one equilibrium state to another:delta G=H- Δ TS, where T is the absolute temperature. The law provides the conceptual basis for redox and group transfer potentials; for example, the Nernst equation or the equilibrium constant, which is delta G=-RTln[Keq], where R is the gas constant. The law allows prediction of the direction in which reversible reactions may proceed, dependent upon reactant and product concentrations, by quantifying and defining free energy for a spontaneous chemical change as a negative value which is independent of the path between the initial and final state. Using transition state theory, the law may be applied to rate constants and their temperature dependencies.

■ **second messenger**
a chemical signal that is generated inside a cell when a hormone (the first messenger) becomes bound to a surface receptor on the outside; e.g. cyclic AMP.

■ **secondary metabolism**
those metabolic pathways, other than those for energy production and for biosynthesis of nucleic acids, proteins, structural components, etc., that are unique to an organism; especially those pathways by which bacteria, moulds and plants synthesise pigments, antibiotics, toxins and other natural products.

■ **secondary structure**
in a protein or a nucleic acid, any repetitive folded pattern that results from the interaction of the corresponding polymeric chains.

■ **sedimentation coefficient**
a physical constant specifying the rate of sedimentation of a particle in a centrifugal field under specified conditions.

■ **sedimentation constant**
a measure of the rate of sedimentation in an analytical ultracentrifuge; v/ω^2x, where v is the rate of sedimentation, is the angular velocity in radians/s and x is the distance from the axis of rotation; usually expressed in Svedberg units (10^{-13}s).

■ **semi-**
half.

■ **semiconservative replication**
duplication of DNA in which the daughter duplex carries one old strand and one new strand.

■ **semipermeable**
the characteristic of allowing only some molecules, usually smaller or uncharged ones, to pass through.

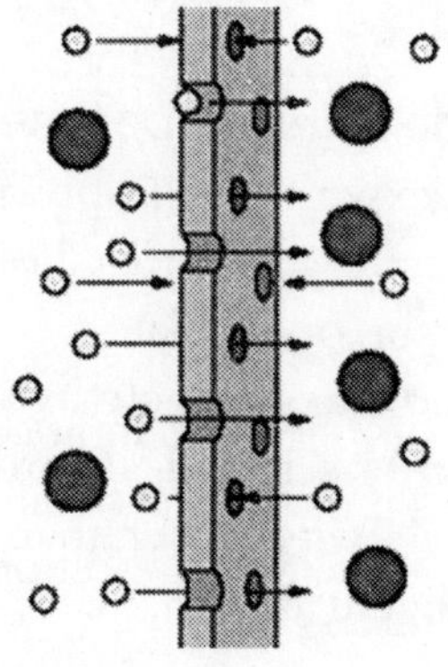

■ **sept-**
seven.

■ **sequencing**
the determination of the base sequence of a homogeneous DNA fragment, the amino acid sequence of a protein, or the glycoside sequence of a complex polysaccharide.

■ **serine proteinase**
a type of peptidase that has at its active site a serine residue.

■ **serotype**
an immunologically defined variant of a micro-organism; one that bears a unique epitope on its surface.

■ **serpin**
serine proteinase inhibitor; one of a group of homologous proteins, e.g. 1-antiproteinase.

■ **serum**
the liquid phase that remains after blood has clotted.

■ **seven-helix motif**
a characteristic of several receptor proteins, that zig-zag seven times as -helices across the plasma membrane.

■ **sex-**

six.

■ **Shine-Dalgarno sequence**

a sequence in an mRNA required for binding prokaryotic ribosomes. The sequence consists of 4-9 purine residues, located 8-13 bp upstream of the initiation codon.

■ **shuffling**

recruitment of the exons that code for a functional domain of one protein in the evolution of a gene for a new protein, to produce hybrids with recognisable motifs, such as kringles and metal-binding fingers. A consequence of this hypothesis is the corollary that introns are as old as structural genes; this is upheld by the intron-early school but opposed by the intron-late school, who point to the absence of introns in prokaryotes and some primitive eukaryotes.

■ **shuttle vector**

a recombinant DNA vector that can be replicated in two or more different host species, making possible the movement of DNA between organisms.

■ **sickle-cell anemia**

a human disease characterised by haemoglobin molecules in which a Glu is replaced by Val. This creates a "sticky patch" when the haemoglobin is deoxygenated, causing haemoglobin molecules to polymerise and deform the red blood cells.

■ **side chain**

the moiety of an amino acid residue in a protein, or of a free amino acid, that is attached to the -carbon and is unique to each amino acid, e.g. the isopropyl group of valine, the benzyl group of phenylalanine. The group attached to C-17 of sterols, mineralocorticoids and bile acids is also called a side chain.

■ **side shelf experiment**

an experimental approach to discovery of a metabolic precursor or regulator, in which candidates for activity are randomly selected according to availability rather than by intelligent design.

■ sigma factor
a subunit of RNA polymerase that recognises specific sites on DNA for initiation of RNA synthesis.

■ sigmoidal
s-shaped; descriptive of the appearance of a graph that shows dependence of rate on substrate concentration for a multimeric enzyme that is accelerated by binding of more than one substrate molecule; also descriptive of non-simple binding of ligands to a non-enzymic protein, e.g. oxygen binding to haemoglobin.

■ signal patch
an area of a protein surface that is recognised by the sorting mechanism of a cell and allows the protein to be directed to its appropriate locus within the cell.

■ signal sequence
an amino-terminal sequence that signals the cellular destination of a newly synthesised protein.

■ signal transduction
the process by which an extracellular signal is converted to an intracellular response.

■ silent mutation
a mutation in a gene that causes no detectable change in the biological characteristics of the gene product.

■ simple diffusion
the unassisted movement of molecules across a membrane to a region of lower concentration.

■ simple protein
a protein yielding only amino acids on hydrolysis.

■ simulated annealing
an energy minimisation technique to refine a model structure, e.g. an X-ray crystallographic or nuclear magnetic resonance model, by simulation of heating it to a high temperature to break all covalent and non-covalent bonds and overcome local energy minima, recalculation of the positions of each atom, then slowly cooling it to allow bonds to re-form between neighbouring atoms.

■ single-copy DNA
a region of the genome whose sequence is present only once per haploid complement.

singlet state

an excited electronic state in which an electron is raised to a higher energy level without reversal of its spin state, and thus is easily able to fall back to its ground state, often with a fluorescent emission.

sirohaeme

a prosthetic group of some redox proteins, consisting of a reduced porphyrin and a Fe_4S_4 iron-sulphur centre.

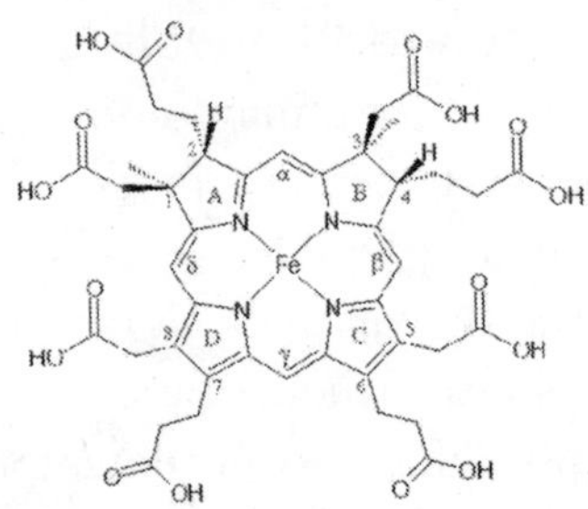

site-directed mutagenesis

a set of methods used to create specific alterations in the sequence of nucleotides in a gene.

site-specific recombination

a type of genetic recombination that occurs only at specific sequences.

slab gel

a thin sheet that acts as a support for electrophoresis, often composed of polyacrylamide or agarose.

small nuclear RNA (snRNA)

any of several small RNA molecules in the nucleus. Most known snRNAs have a role in the splicing reactions that remove introns from mRNA, tRNA, and rRNA molecules.

soap

a sodium, potassium or other soluble salt of a fatty acid.

solenoidal model

a representation of the organisation of chromatin in which adjacent nucleosomes in the linear DNA sequence are wound into a 36nm diameter helix that contains six nucleosomes per turn.

somatic cells

body cells. All the cells except the germ-line cells.

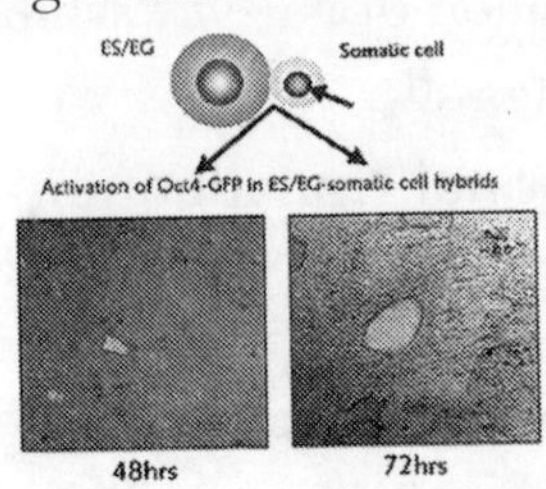

■ somatic hybrid

the fusion product of two different types of non-gamete cells.

■ sorting

the translocation to their appropriate loci within a cell of newly synthesised or endocytosed proteins and nucleic acids.

■ SOS response

in bacteria, a coordinated induction of a variety of genes as a response to high levels of DNA damage.

■ SOS system

a set of DNA repair enzymes and regulatory proteins that regulate their synthesis so that maximum syn thesis occurs when the DNA is damaged.

■ southern blotting

a method for detecting a specific DNA restriction fragment, developed by Edward Southern. DNA from a gel electrophoresis pattern is blotted onto nitrocellulose paper; then the DNA is denatured and fixed on the paper. Subsequently the pattern of specific sequences in the Southern blot can be determined by hybridisation to a suitable probe and autoradiography. A Northern blot is similar, except that RNA is blotted instead onto the nitrocellulose paper.

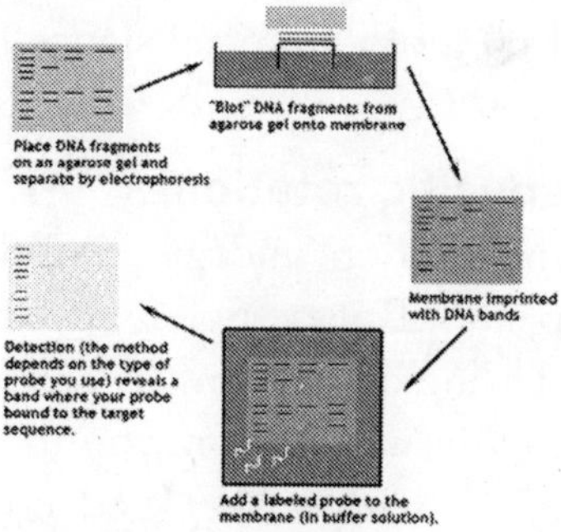

■ spacer DNA

DNA that does not yet have a recognised function.

■ special pairs

antenna molecules which have associated with them proteins required for the initial separation of charge during photosynthesis.

■ specific activity

the number of μmol of a substrate transformed by an enzyme preparation per minute per milligram of protein at 25°C. Specific activity is used to estimate enzyme purity during a purification schaeme.

■ **specific base**
in chemistry, a hydroxyl ion that can participate in catalysis.

■ **specific heat**
the amount of energy needed to raise the temperature of 1.00 g of a pure substance by 1.00°C.

■ **specific rotation**
the rotation, in degrees, of the plane of plane-polarised light (D-line of sodium) by an optically active compound at 25°C, with a specified concentration and light path.

■ **specificity**
the ability of an enzyme or receptor to discriminate among competing substrates or ligands.

■ **sphingolipid**
an amphipathic lipid with a sphingosine backbone to which are attached a long-chain fatty acid (at the nitrogen) and a polar alcohol.

Sphingolipid Metabolic pathways

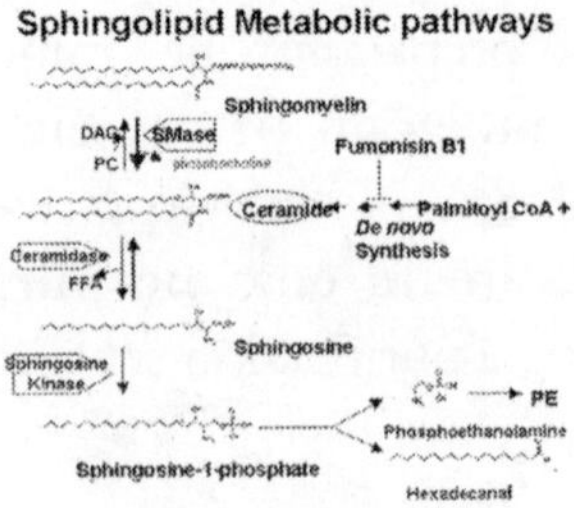

■ **spin lattice**
the magnetic environment that surrounds the nucleus or electron under observation in nuclear magnetic resonance or electron spin resonance studies.

■ **splicing**
see RNA splicing

■ **sporulation**
formation from vegetative cells of metabolically inactive cells that can resist extreme environmental conditions.

■ **ß conformation**
an extended, zigzag arrangement of a polypeptide chain which can be parallel or antiparallel. One of the common secondary structure in proteins.

■ **ß oxidation**
oxidative degradation of fatty acids into acetyl-CoA by successive oxidation, reduction and hydrolytic reactions at the ß-carbon atom.

■ **ß2-microglobulin**
11.8 kDa protein.

■ stacking energy
the energy of interaction that favors the face-to-face packing of purine and pyrimidine base pairs.

■ standard free-energy change (ΔGº)
the free-energy change for a reaction occurring under a set of standard conditions: temperature, 298 K; pressure, 1 atm; and all solutes at 1 M concentration. ΔGº' denotes the standard free-energy change at pH 7.0.

■ standard reduction potential (Eo')
the electromotive force exhibited at an electrode by 1 M concentrations of a reducing agent and its oxidised form at 25ºC and pH 7.0. It describes the relative tendency of the reducing agent to lose electrons.

■ steady state
in enzyme-kinetic analysis, the time interval when the rate of reaction is approximately constant with time. The term is also used to describe the state of a living cell where the concentrations of many molecules are approximately constant because of a balancing between their rate of synthesis and breakdown.

■ stem cells
the precursor, self-regenerating cells in bone marrow that give rise to differentiated blood cells such as erythrocytes and lymphocytes.

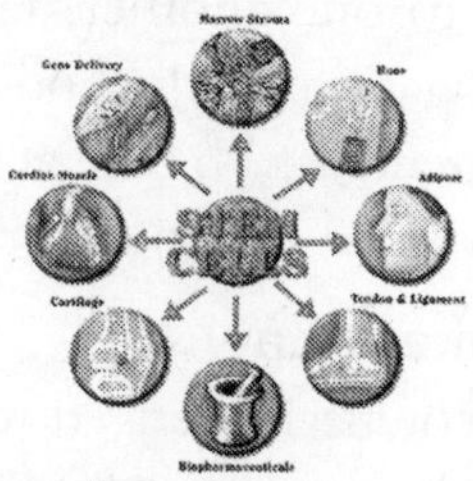

■ stereoisomers
compounds that have the same molecular formula and similar structural formulas, but differ in the arrangement around one chiral center.

■ steroids
compounds that are derivatives of a tetracyclic structure composed of a cyclopentane ring fused to a substituted phenanthrene nucleus.

■ sterol
a crystalline alcohol that incorporates a characteristic perhydrocyclopentanophenanthrene ring system and a branched

hydrocarbon side chain, e.g. cholesterol, ergosterol.

■ sticky ends (cohesive ends)

two DNA ends in the same DNA molecule, or in different molecules, with short overhanging single-stranded segments that are complementary to one another, facilitating ligation of the ends. Usually generated by the action of restriction enzymes.

■ Stokes' Law

a theoretically derived relationship between the viscosity coefficient, eta , the frictional coefficient, f, and the radius of a macromolecule (the Stokes radius) r: f=6r eta. This calculated f may be compared with that determined experimentally from the relationship between the diffusion coefficient, D, the Boltzmann constant, k, and the absolute temperature, T: D=kT/f. The deviation (usually an increase) of the measured f from the calculated f is a measure of the solvation of the macromolecule or its deviation from a spherical shape.

■ stop codons

UAA, UAG, and UGA. In protein synthesis these codons signal the termination of a polypeptide chain. AKA termination codons or nonsense codons.

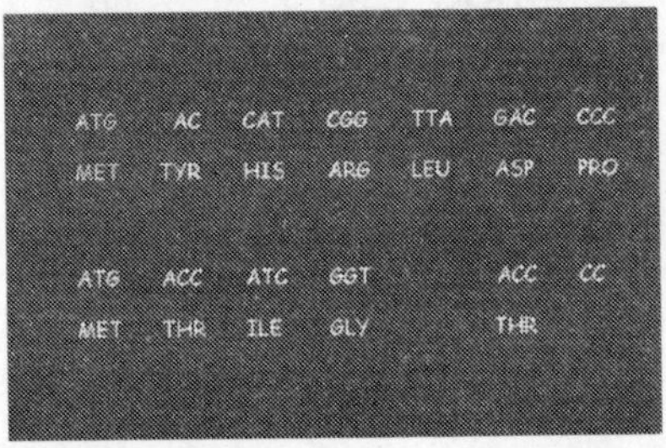

■ stroma

the space and aqueous solution enclosed within the inner membrane of a chloroplast.

■ structural domain

an element of protein tertiary structure that forms an independent folding unit.

■ structural gene

a gene encoding the amino acid sequence of a polypeptide chain.

■ structural protein

a protein that serves a structural function.

■ sub-

under, beneath.

■ substitution mutation

a mutation caused by the replacement of one base by another.

■ Substrate

a molecule that is acted upon, and chemically changed, by an enzyme.

■ substrate channelling

the sequential action of a multienzyme complex upon a substrate that prevents the dissociation of intermediate products.

■ substrate phage

a method for identification of the substrate specificity of a proteinase by its ability to select a phage that expresses a cleaveable polypeptide, which is one of a library of random amino acid sequences fused to a protein that tethers the phage to a solid support. Selection of the phage is by cleavage of the polypeptide. Several cycles of binding, proteolysis and phage propagation select the phage that expresses the scissile bond.

■ substrate-level phosphorylation

phosphorylation of ADP or some other nucleoside 5'-diphosphate coupled to the dehydrogenation of an organic substrate; independent of the electron transport chain.

■ subunit

individual polypeptide chains in a protein.

■ suicide inhibitor

a relatively inert molecule that is transformed by an enzyme, at its active site, into a reactive substance that irreversibly inactivates the enzyme.

■ sulphur cycle

the passage of sulphur from organic to inorganic forms, from substituents of cells to minerals, via the following steps: (1) uptake of sulphate from soil by organisms and conversion into organic sulphur, or reduction to the gas dimethylsulphide (e.g. by algae and cyanobacteria), (2) photochemical oxidation to methanesulphonic acid in the atmosphere, (3) return to the soil as it falls to earth in rain and snow, and (4) oxidation to sulphate by soil bacteria.

SIMPLIFIED SULPHUR CYCLE FOR MINERAL DEPOSITS

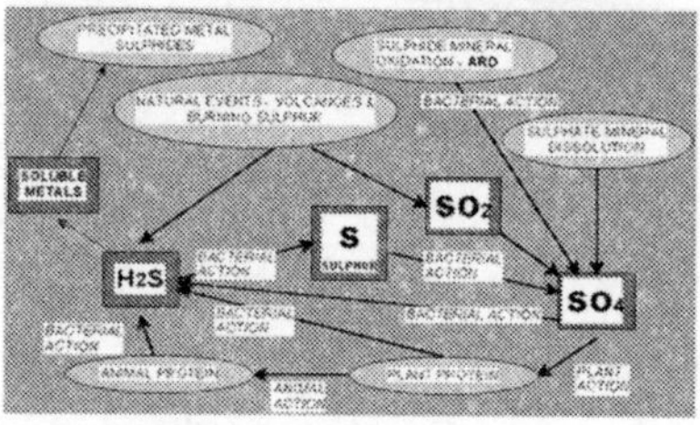

■ **super-**
above, in addition, over.

■ **supercoiled DNA**
supertwisted, covalently-closed duplex DNA.

■ **superhelix**
the twisted axis of a fibrous polymer, such as double-stranded DNA, myosin tails or collagen.

■ **suppressor gene**
a gene that can reverse the phenotype of a mutation in another gene.

■ **suppressor mutation**
a mutation that totally or partially restores a function lost in a previous mutation. Must be located at a site different from the site of the first mutation.

■ **supra-**
above, on the upper side.

■ **Svedberg unit (S)**
the unit used to express the sedimentation constant (S = 10^{-13} sec). The sedimentation constant S is proportional to the rate of sedimentation of a molecule in a given centrifugal field and is related to the size and shape of the molecule.

■ **switching**
during development, the turning off of one gene and the turning on of a related one; e.g. during faetal development, the change from synthesis of gama-globin to synthesis of beta-globin.

■ **symbiosis**
two (or more) organisms that are mutually interdependent and live in physical association.

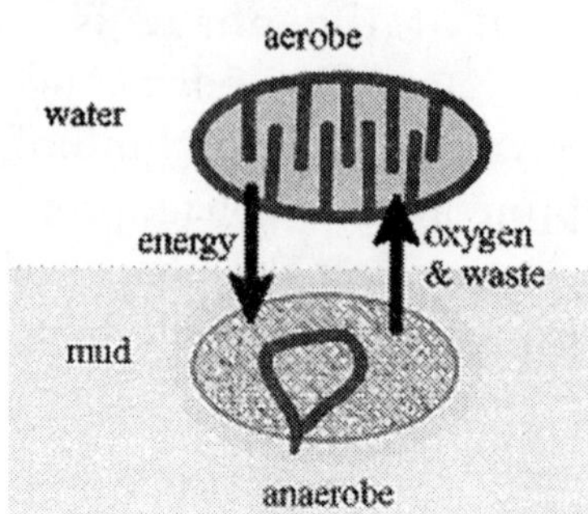

■ **symport**

cotransport of solutes across a membrane in the same direction.

■ **syn-**

together, with.

■ **synapse**

the chemical connection for communication between two nerve cells or between a nerve cell and a target cell such as a muscle cell.

■ **synapsis**

the pairing of homologous chromosomes, seen during the first meiotic prophase.

■ **synthase**

a lyase (acting in the reverse of the reaction) that adds one substrate across the double bond of another, e.g. delta-aminolaevulinic acid synthase, hydroxymethyl glutaryl-CoA synthase.

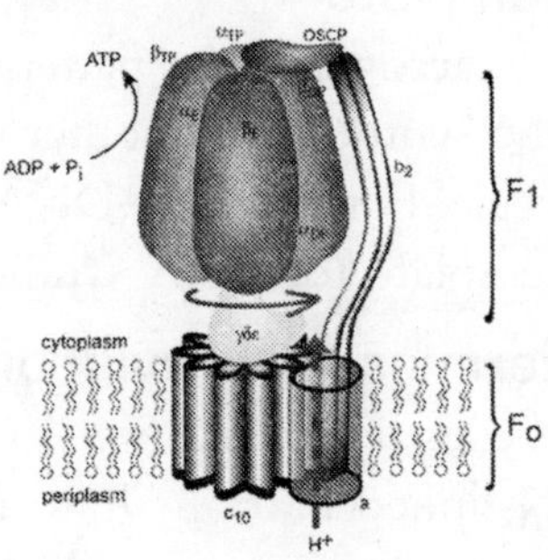

■ **synthases**

enzymes that catalyse reactions in which no NTP is required as an energy source.

■ **synthetase**

an enzyme, regardless of its mechanism, that couples two substrates by a carbon-carbon, carbon-oxygen, carbon-sulphur or carbon-nitrogen bond that is driven by the hydrolysis of a phosphoanhydride bond, e.g. amino acyl-tRNA synthetase, fatty acyl-CoA synthetase. Those enzymes that catalyse such syntheses and do not use the energy of a phosphoanhydride bond are synthases (despite their frequently being misnamed).

■ **system**

an isolated collection of matter in thermodynamics. All other matter in the universe apart from the system is called the surroundings.

■ **tandem array**

a DNA structure in which a gene and associated sequences are repeated in an immediately adjacent position.

■ tandem duplication

a duplication in which the repeated regions are immediately adjacent to one another.

■ tannin

a product of plant origin that converts hides into leather; often polyphenols and aldehydes that cross-link collagen chains and render them insoluble and unreactive to enzymes that would degrade and putrefy untreated hides. Chromic acid salts are 'tanning agents' but not tannins.

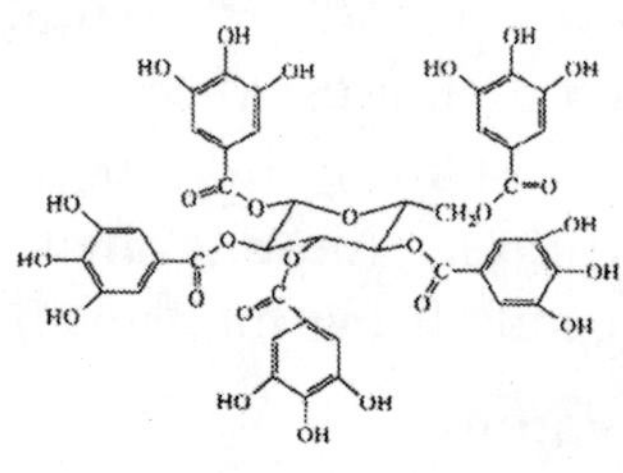

■ target cell

a cell that is stimulated by a particular hormone or neurotransmitter.

■ TCA cycle

see tricarboxylic acid cycle

■ teichoic acid

a linear polymer present in Gram-positive bacterial capsules, cell walls and membranes, characterised by polyols (glycerol, ribitol, sugars), sometimes derivatised as D-alanine esters, and linked together through phosphodiester bonds.

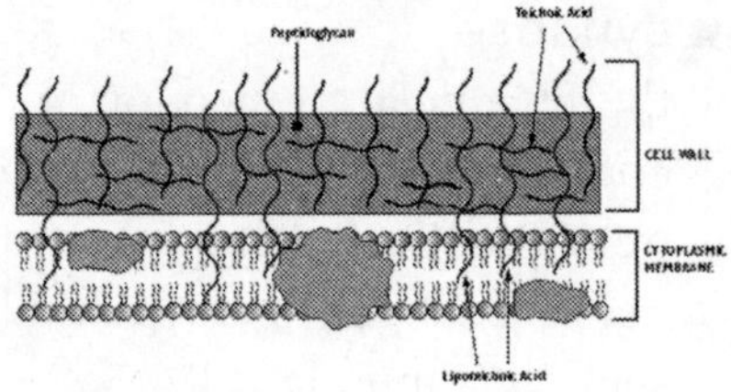

■ telomere

a specialised and not well understood nucleic acid structure found at the ends of linear eukaryotic chromosomes.

■ temperate phage

a phage whose DNA may be incorporated into the host-cell genome without being expressed. Contrast with a virulent phage, which destroys the host cell.

■ template

a macromolecular pattern for the synthesis of another molecule. For example, DNA is a template for RNA synthesis.

■ terminal glycosylation

the final modifications of the N-linked carbohydrate moieties of a protein that are effected in the Golgi apparatus.

■ terminal transferase
an enzyme that catalyses the addition of nucleotide residues of a single kind to the 3' end of DNA chains. For example polyA synthetase.

■ termination codons
stop codons.

■ termination factors
protein factors of the cytoplasm required in releasing a completed polypeptide chain from a ribosome. AKA release factors.

■ termination sequence
a DNA sequence that appears at the end of a transcriptional unit and signals the end of transcription.

■ terpenes
hydrocarbons or hydrocarbon derivatives constructed from recurring isoprene units in many plant species. Among other things, these molecules are responsible for some of the scents of plant products, for example, the smell of pine needles.

■ tertiary structure
the three-dimensional conformation of a protein in its native folded state.

■ tetra-
four.

■ tetrad
a four-stranded DNA structure, e.g. the i-tetrad, composed of parallel and anti-parallel poly (C) sequences in which all strands are equivalent and each C-C base pair shares a proton.

■ tetrahydrobiopterin
the fully reduced coenzyme form of biopterin.

■ tetrahydrofolate
the reduced, active coenzyme form of the vitamin folate which functions as an antioxidant.

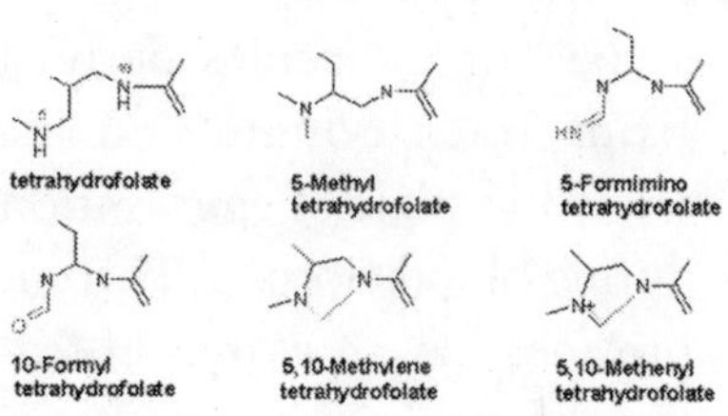

■ tetramer
structure resulting from the association of four subunits.

■ thiamine pyrophosphate
the active coenzyme form of vitamin B1 which functions in aldehyde transfer reactions.

■ thio-
sulphur.

■ thioester
an ester of a carboxylic acid with a thiol or mercaptan.

■ thiosulphate shunt
a pathway for anaerobic metabolism of sulphide by bacteria, e.g. in marine sediments; thiosulphate ($S_2O_3^{2-}$) is an intermediate between sulphide (S_2-) and sulphate (SO_4^{2-}) in reduction to the former, oxidation to the latter and disproportionation to both.

■ thrombocytes
platelets.

■ thromboxanes
a class of molecules derived from arachidonate and involved in platelet aggregation during blood clotting. Thromboxanes have a common feature of a 6-membered, ether-containing ring. [a thromboxane]

■ thylakoid
closed cisterna, or disc, formed by the pigment-bearing internal membranes of chloroplasts.

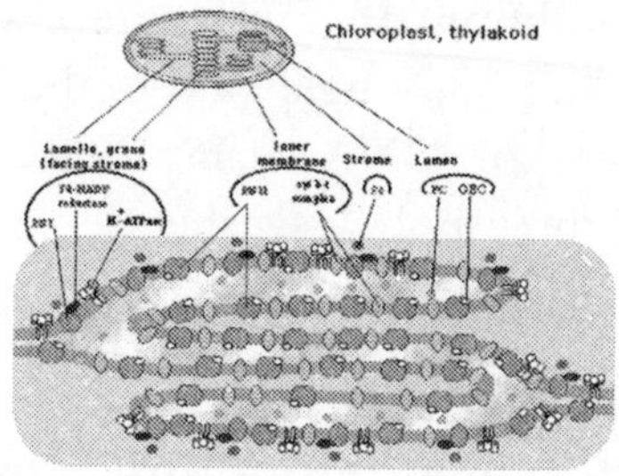

■ thymidine
one of the four nucleosides found in DNA.

■ thymine
a pyrimidine base found in DNA.

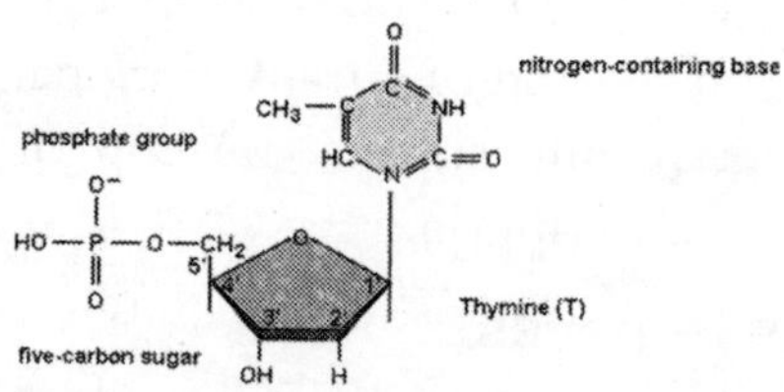

■ thymine dimer
pyrimidine dimer.

■ tight junction
the seal that closes the gap between adjacent epithelial cells to ensure closure of a vascular space. Transport through a tight junction is

paracellular, in contrast to transcellular transport through cells.

■ **tissue culture**
method by which cells derived from multicellular organisms are grown in liquid media.

■ **titration curve**
a graphical representation of the protonic dissociation of a compound, e.g. pH against equivalents of alkali added to the acid form of the compound.

■ **tocopherols**
forms of vitamin E.

■ **tomy**
cut.

■ **topoisomerases**
enzymes that introduce positive or negative supercoils in closed, circular duplex DNA.

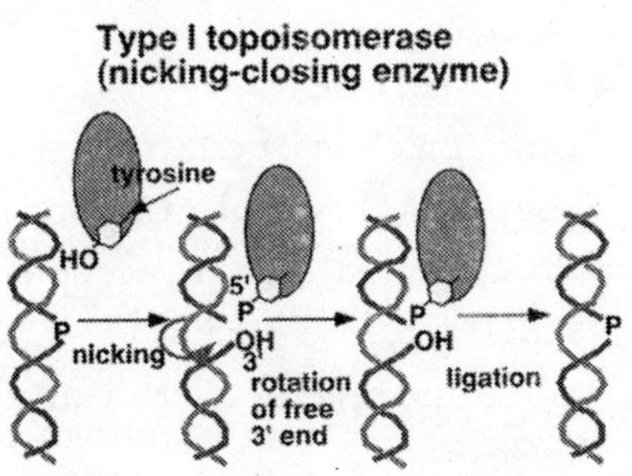

■ **toxins**
proteins produced by some organisms and toxic to some other species.

■ **trace element**
a chemical element (such as Se) required by an organism in only trace amounts.

■ **trans-**
across, beyond.

■ **transamidation**
the substitution of one amino compound for another in an amide linkage, e.g. the cross-linking of fibrin by the displacement of NH_3 from glutamine residues by an epsilon-amino group of a lysine residue to form an isopeptide bond.

■ **transaminases**
aminotransferases.

■ **transamination**
the concurrent amination and reduction of a carbonyl of one compound as another is deaminated and oxidised; especially the conversion of one alpha-amino acid into its corresponding alpha -oxo acid as another alpha -oxo acid is converted into its corresponding

alpha -amino acid, mediated by the pyridoxal phosphate prosthetic group of a transaminase (aminotransferase).

transcription
the enzymatic process whereby the genetic information contained in one strand of DNA is synthesised into a complementary sequence of bases in an mRNA chain.

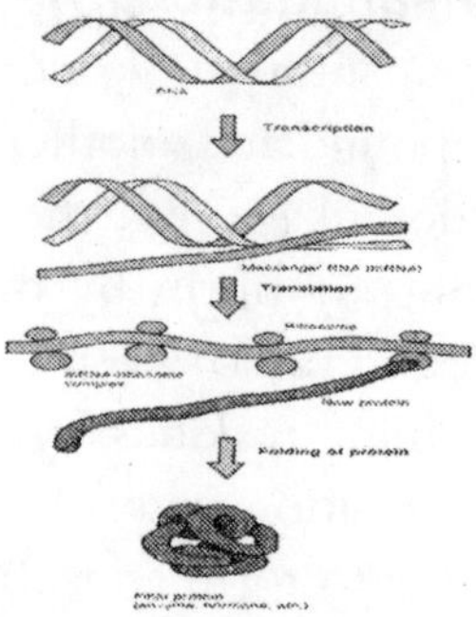

transcription unit
the sequence from the transcriptional start site to the site of termination, inclusive of introns.

transcriptional control
the regulation of the rate of a protein's synthesis by regulation of the formation of its mRNA.

transduction
(1) The conversion of energy from one form to another (ATP synthesis by a chemiosmotic mechanism). (2) The conversion of information from one form to another (epinephrine outside a liver cell causing cAMP inside the cell). (3) The transfer of genetic information from one cell to another by means of a viral vector.

Transfection
an artificial process of infecting cells with naked viral DNA.

transfer RNA (tRNA)
the class of RNA molecules which are ~75 nucleotides in size, having a cloverleaf structure and many modified bases.

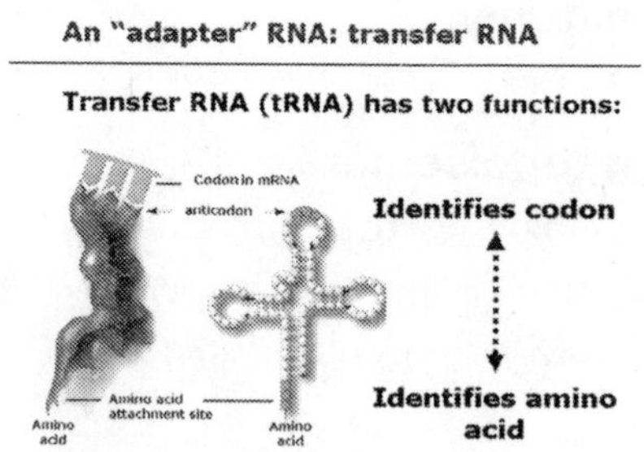

transferase
one of a class of enzymes that transfer a chemical group from donor substrates to acceptor substrates, e.g. a kinase, a phosphorylase, a transaminase.

■ transferrin

a 90 kDa protein responsible for the transport of iron and uptake by the liver.

■ transformation

genetic exchange in bacteria that is mediated via purified DNA. In somatic cell genetics the term is also used to indicate the conversion of a normal cell to one that grows like a cancer cell.

■ transgenic

descriptive of an animal that has developed in a surrogate mother from a fertilised ovum that had been injected with a recombinant DNA gene; contrasted with congenic, meaning of the same genetic origin.

■ transition state

an activated form of a molecule in which the molecule has undergone a partial chemical reaction and has bond characteristics of both reactant and product.

■ translation

the process in which the genetic information present in a mRNA molecule is translated into a sequence of amino acids (protein synthesis).

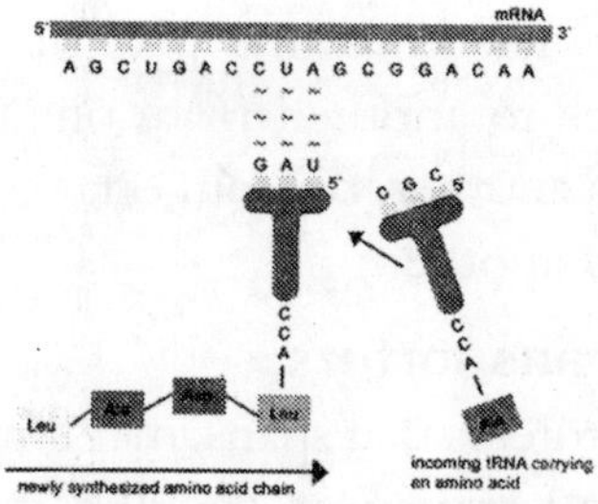

■ translational control

the regulation of the rate of a protein's synthesis by regulation of the rate of translocation of the ribosome.

■ translational repressor

a repressor that blocks translation of a mRNA.

■ transmethylation

the metabolic transfer of preformed methyl groups from one acceptor to another, e.g. from S-adenosylmethionine to guanidinoacetate to form creatine.

■ transpiration

passage of water from the roots of a plant to the atmosphere *via* the vascular system and the stomata of the leaves.

■ transport protein

a protein whose primary function is to transport a substance from one part of the cell to another, from one cell to another, or from one tissue to another.

■ transporters

proteins that span a membrane and transport specific molecules across the membrane.

■ transposition

the movement of a gene or set of genes (transposon) from one site in the genome to another.

■ transposon

one of a class of genes that are capable of moving spontaneously from one chromosome to another, or from one position to another in the same chromosome; also known as jumping genes or transposable elements.

■ tri-

three.

■ triacyiglycerol

an ester of glycerol with three molecules of fatty acid; also called a triglyceride or neutral fat.

■ tricarboxylic acid cycle (TCA)

the cyclical process whereby acetate is completely oxidised to CO_2 and water, and electrons are transferred to NAD^+ and flavine. The TCA cycle is localised to the mitochondria in eukaryotic cells and to the plasma membrane in prokaryotic cells. Also called the Krebs or citric acid cycle.

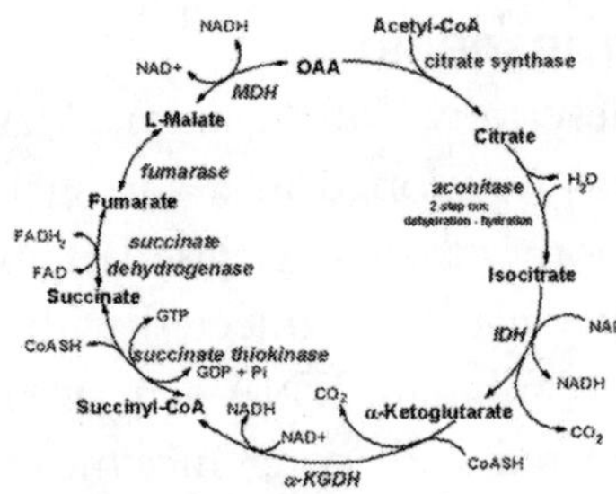

■ triose

a simple sugar with a backbone containing three carbon atoms.

■ triple helix

the tertiary structure of collagen that twists three polypeptide chains around themselves; also a triple-stranded DNA structure that involves Hoogstein base pairing between B-DNA and a third DNA strand that occupies the major groove.

■ **tRNA**

transfer RNA.

■ **Trojan horse inhibitor**

affinity labelling reagent.

■ **tropic hormone (tropin)**

a peptide hormone that stimulates a specific target gland to secrete its hormone. For example, lutropin (LH) produced by the anterior pituitary stimulates production of progesteron by the corpus luteum.

■ **trypsin**

a proteolytic enzyme that cleaves (cuts) peptide chains next to the basic amino acids arginine and Iysine.

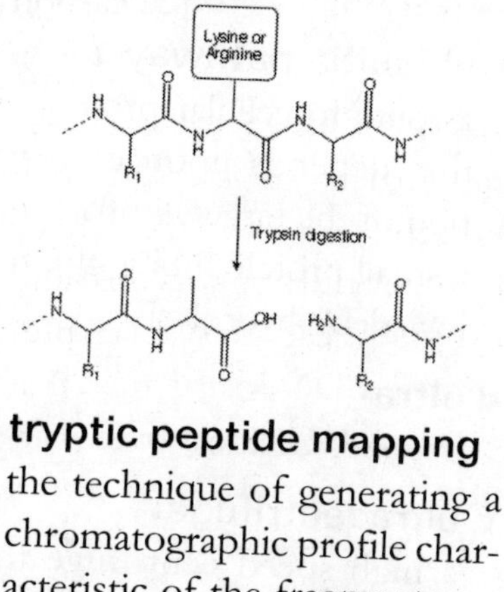

■ **tryptic peptide mapping**

the technique of generating a chromatographic profile characteristic of the fragments resulting from trypsin enzyme cleavage of the protein.

■ **tumorigenesis**

the mechanism of tumor formation.

■ **tumour**

a neoplasm; a growth of new tissue, sometimes malignant.

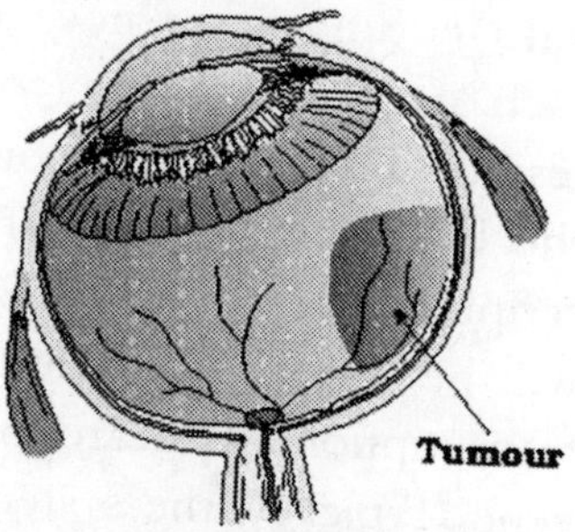

■ **turnover number**

the number of times an enzyme molecule transforms a substrate molecule per sec, under conditions giving maximal activity at substrate concentrations that are saturating.

■ **twintron**

an intron within an intron; the internal intron is excised before the external intron is spliced out.

■ **two-hybrid system**

an assay for detecting in vivo protein-protein interactions, which depends upon expression of a reporter gene, usually in yeast. As transcription of eukaryotic enzymes are regulated by transcriptional activators having distinct DNA-binding and transcriptional activator domains, the coding sequences for each of these domains can be fused with coding sequences for each of the two proteins to be tested; if the proteins bind, the DNA-binding and transcriptional activator domains will be reunited, permitting transcription of the reporter gene. Typically, the system is used to detect a binding partner of a known protein (encoded in the bait vector as a chimera with the DNA binding domain of the transcription factor) to a candidate partner (encoded in the prey vector as a chimera with the activation domain of the transcription factor), which is the homologue of a known binding partner or is encoded by a cDNA library.

■ **Tyndall effect**

the phenomenon of light scattering that is dependent upon the wavelength of light. The intensity of light scattered by small isotropic particles is inversely proportional to the fourth power of the wavelength.

■ **tyrosine kinase**

an enzymic activity associated with the cytoplasmic domain of several growth factor receptors. Two such molecules, when liganded to an effector molecule by their extracellular domains, phosphorylate tyrosine residues on each other; this increases their kinase activities so that they are then able to phosphorylate other cytoplasmic proteins, thus continuing the signal transduction pathway initiated by effector binding to the receptor.

■ **ubiquitin pathway**

a route to cellular proteolysis that depends upon conjugation of the targeted protein to a small protein, ubiquitin, followed by hydrolysis.

■ **ultra-**

beyond, besides, over.

■ **ultracentrifuge**

a high-speed centrifuge that can attain speeds up to 60,000

rpm and centrifugal fields of 500,000 times gravity. Useful for characterising and/or separating macromolecules.

ultrafiltration

a separation procedure in which a solution is forced through a membrane with a pore size that is selected to retain macromolecules of a certain size and to pass smaller ones.

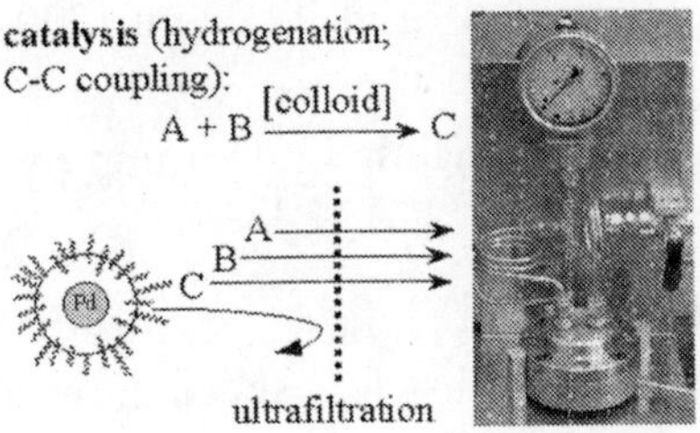

ultraviolet (UV) radiation

electromagnetic radiation in the region of 200 to 400 nm.

uncompetitive inhibition

a form of enzyme inhibition in which the inhibitor binds to the enzyme-substrate complex, resulting in decreases in the Km and Vmax values.

uncoupling agent

a substance that uncouples phosphorylation of ADP from electron transfer by disrupting the pH gradient; for example, 2,4-dinitrophenol.

uni-

one.

uniport

a transport system that carries only one solute. Contrast with cotransport.

unnatural amino acid mutagenesis

a method for replacement of an amino acid residue of a protein with a non-naturally occurring amino acid. The mRNA that encodes the protein is modified to encode a nonsense suppressor codon. For example, mRNA is modified to use the UAG codon; this mRNA, together with a nonsense suppressor tRNA with the anticodon CUA, and with an unnatural amino acid at its 3'-end, is introduced into an expression system, possibly a Xenopus oocyte.

unsaturated fatty acid

a fatty acid containing one or more double bonds.

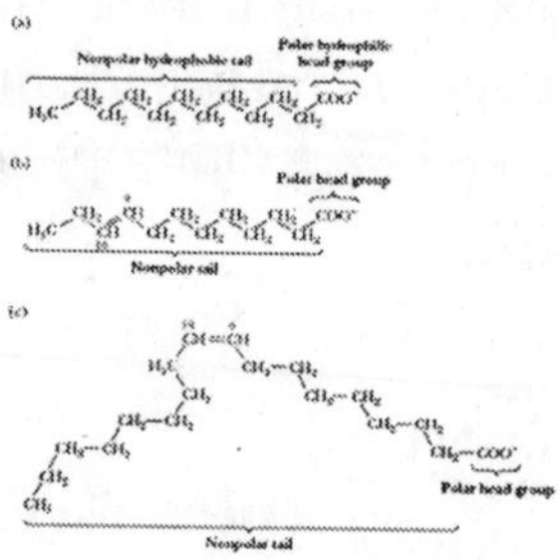

■ **unwinding proteins**
proteins that help to unwind double-stranded DNA during DNA replication.

■ **upstream**
in a polynucleotide chain, towards the 5'-end.

■ **urea cycle**
a metabolic pathway in the liver that leads to the synthesis of urea from amino groups and CO_2. The function of the pathway is to convert the ammonia resulting from catabolism to a nontoxic form, which is then secreted.

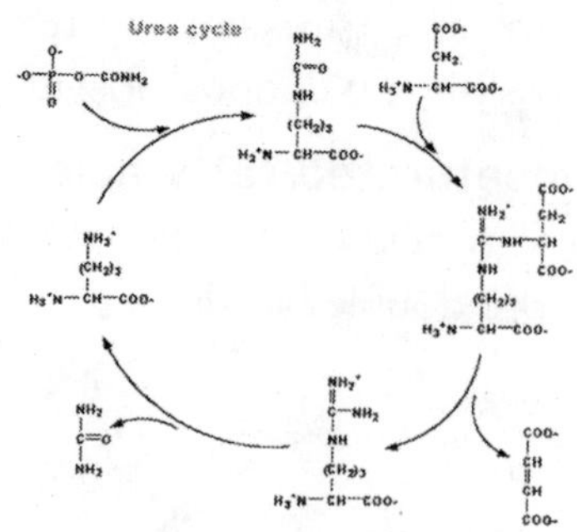

■ **ureotelic**
descriptive of an organism that synthesises and excretes urea as the final product of nitrogen metabolism.

■ **uronic acid**
a sugar derivative in which the hydroxymethyl group is oxidised to a carboxyl group.

■ **UV irradiation**
electromagnetic radiation with a wavelength shorter than that of visible light (200-390 nm). Causes damage to DNA (mainly by forming pyrimidine dimers).

■ **vacuole**
a large vesicle, especially in a plant cell.

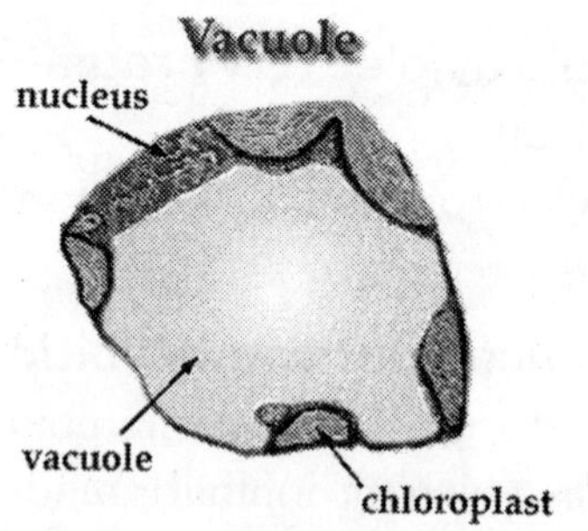

■ **Van den Bergh reaction**
a method for colorimetric estimation of serum bilirubin by coupling with diazotised sulphanilic acid. The direct van den Bergh reaction, in

aqueous medium, yields the 'direct' bilirubin, i.e. the amount of water-soluble bilirubin glucuronide conjugates. Upon prior addition of methanol to solubilise free bilirubin, the total bilirubin is found, and by this indirect van den Bergh reaction the 'indirect' bilirubin, i.e. the amount of unconjugated bilirubin, may be calculated as the difference between the total and direct levels.

■ Van der Waals forces

refers to the combined effect of two types of interactions, one attractive and one repulsive. The attractive forces are due to favorable interactions among the induced instantaneous dipole moments that arise from fluctuations in the electron charge densities of neighboring non-bonded atoms. Repulsive forces arise when noncovalently bonded atoms come too close together.

■ variable region

the sequences of immunoglobulin light or heavy chains that show variation from one particular antibody to another, and are responsible for the binding of antigen.

■ vasodilation

the increase in diameter of small blood vessels, especially in the kidney, that lowers blood pressure.

■ vector

a DNA molecule known to replicate autonomously in a host cell, to which a segment of DNA may be spliced to allow its replication. The current types are plasmids, cosmids, YAC's or a temperate-phage DNA.

■ vesicle

a membrane-limited sac in the cytoplasm of a cell.

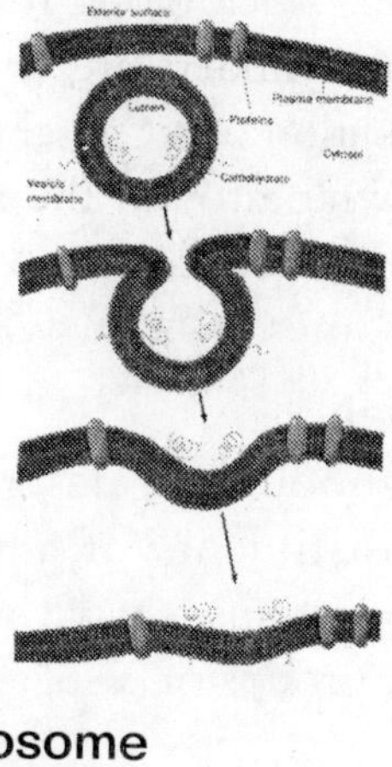

■ vesosome

an aggregate of smaller particles packaged into a larger

membrane-bound structure. The aggregates are formed by attachment of biotin to individual liposomes or other small particles, then cross-linked with streptavidin. The larger aggregates thus formed are partially broken into appropriate sizes by extrusion. In parallel, biotin-labeled films of phosphatidylserine or phosphatidylethanolamine in the presence of Ca^{2+} spontaneously roll up into jelly roll-like cochleate cylinders. The biotin labels of the cylinders are saturated with streptavidin and mixed with the aggregates, which adhere to the cylinders through their remaining open biotin residues. As the cylinders are uncoiled by addition of a Ca^{2+} chelator, they wrap around the adhering aggregates to form vesosomes.

■ vicinal

'Neighbouring'; descriptive of substituents of adjacent carbon atoms, e.g. vicinal hydroxy groups of a sugar.

■ vinca alkaloid

also called vinblastine or vincristine; an inhibitor of microtubule polymerisation and an anti-mitotic agent.

■ viral vector

a viral DNA altered so that it can act as a vector for recombinant DNA.

■ virion

the extracellular form of a virus, complete with its capsid.

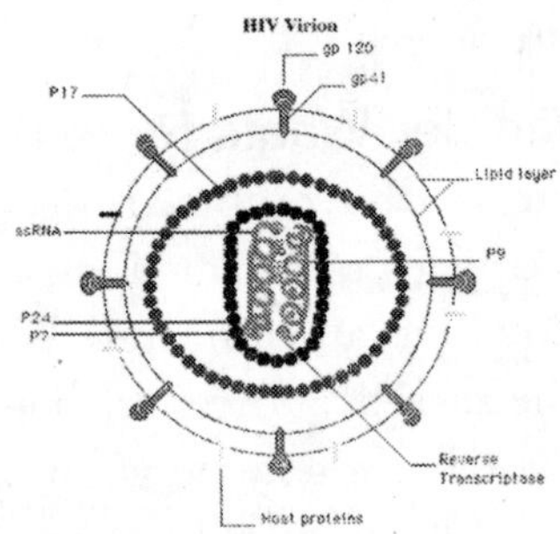

■ Viroids

pathogenic agents, mostly of plants, that consist of short (usually circular) RNA molecules.

■ virus

an obbligate intracellular parasite. A particle composed of a nucleic acid-protein complex that infects an intact host cell and uses that host cell's machinery for its replication.

Can contain either a DNA or RNA genome.

■ vitamin

a compound required in the diet in minute amounts; fat-soluble if soluble in organic solvents (vitamins A, D and E); water-soluble if soluble in aqueous solutions (ascorbic acid, biotin, cobalamin, folic acid, nicotinamide, pantothenic acid, pyridoxine, riboflavin and thiamin).

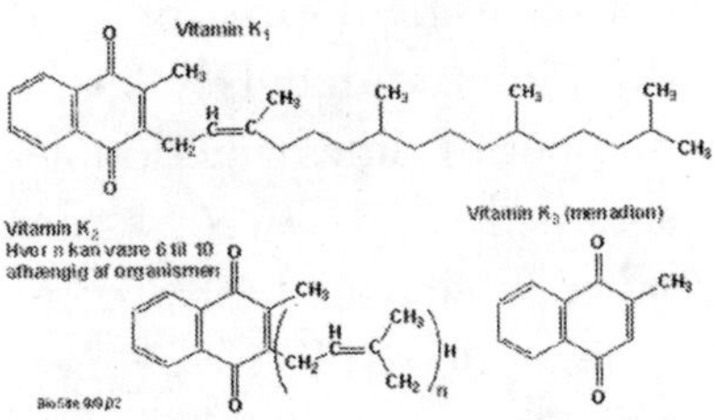

■ Vmax

the maximum velocity of an enzymatic reaction when the binding site is saturated with substrate.

■ Walden inversion

a change of the configuration at an asymmetrical carbon atom due to the entry from one side of the centre of an attacking group, simultaneously with the departure from the other side of a leaving group.

■ Watson-Crick base pairs

the type of hydrogen-bonded base pairs found in DNA, or comparable base pairs found in RNA. The base pairs are A-T, G-C, and A-U.

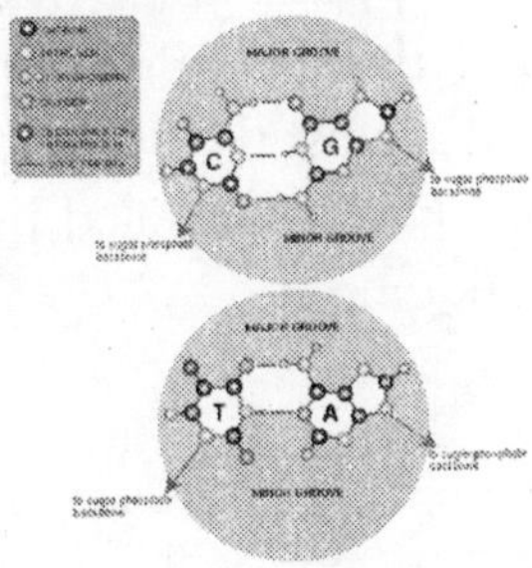

■ wax

the product of esterification of a fatty acid with an alcohol other than glycerol, usually a long-chain alcohol.

■ weaning

the removal of a young mammal from a diet of its mother's milk.

■ western blot

similar in principle to a Southern blot, but where the species adsorbed to the nitrocellulose filter is a protein, and the detection makes use of specific antibodies.

■ western blotting (immunoblotting)

a technique to detect specific proteins. A mixture of proteins is separated by polyacrylamide-gel electrophoresis, blotted on to a plastic sheet (e.g. of nitrocellulose) and then exposed to a radiolabelled immunoglobulin directed against the desired protein, which is revealed by autoradiography.

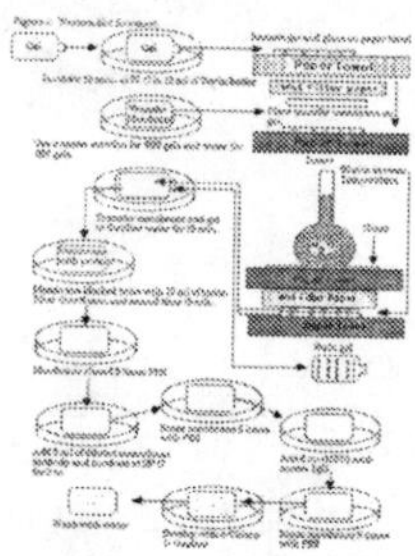

■ white muscle (fast-twitch muscle)

a poorly vasculariSed form of skeletal muscle characterised by little myoglobin and few mitochondria; supplied with energy mainly by glycolysis.

■ wild type

in genetics, the unmutated allele or the natural allele that predominates over other variants in a population.

■ wild-type gene

the form of a gene (allele) normally found in nature.

■ wobble

the presumed slackness of structural requirements for complementarily in fitting codon to anticodon that permits several bases in the third position of the codon to pair with a particular base in the first position of the anticodon.

■ writhing number (W)

a topological property of double-stranded DNA; the number of times one double-stranded segment crosses another in circular DNA, commonly referred to as a supercoil; equal to the linking number minus the twisting number (W=L-T).

■ xenobiotic

a non-biological compound, often one that an organism must eliminate or neutralise by some detoxification strategy.

■ xenopus oocyte

an egg of the African clawed frog, Xenopus laevis; a favoured heterologous expression system that, because of its large size, can be microinjected with cDNA or mRNA and monitored for synthesis of a specific protein.

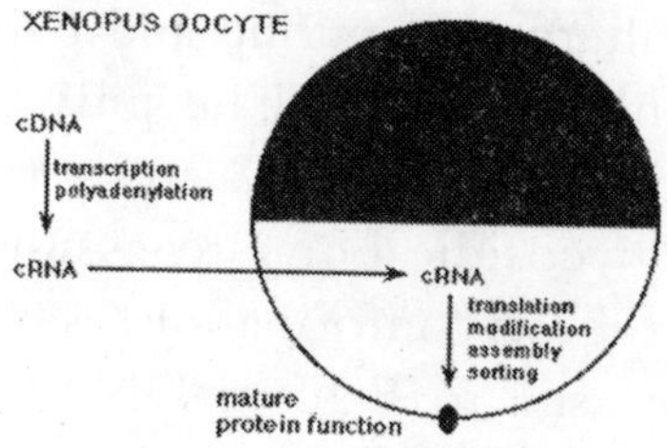

■ X-ray crystallography

a method for structural analysis of solids that have a repeating unit; dependent upon the ability of electrons to scatter X-rays, and capable of locating in space atoms larger than hydrogen in proteins and nucleic acids as well as in less complex compounds.

■ YAC

yeast artificial chromosome.

■ yeast artificial chromosome (YAC)

a cloning vector that can accept a relatively large fragment of foreign DNA, up to 1Mb, in yeast cells.

■ yellow enzyme

one of the two first-discovered flavoenzymes. The 'new' yellow enzyme (das neue gelbe Ferment) catalyses the reduction of molecular oxygen or Methylene Blue by glucose 6-phosphate, as is now understood via reduction by NADPH, and has an FAD prosthetic group. The first-discovered 'old' yellow enzyme (das alte gelbe Ferment) has an FMN prosthetic group but has no recognised biological function.

■ Yin-Yang hypothesis

the now-abandoned theory that bidirectional cellular processes, i.e. those that involve two individual pathways (rather than one that may be 'on' or 'off'), e.g. proliferation/ contact inhibition, glycogen synthesis/ breakdown, are regulated by a balance of the opposed actions of cyclic AMP and cyclic GMP.

■ Y-joint

a DNA heteroduplex structure in which one strand is extended by a non-homologous internal sequence. The loop of the longer strand is base-paired with itself in a hairpin structure. In an open Y-joint the shorter, continuous strand is nicked at the joint; in a closed Y-joint the continuous strand is unnicked.

■ ylid

a compound in which adjacent, covalently-bonded atoms, both having an electronic octet, have opposite charges.

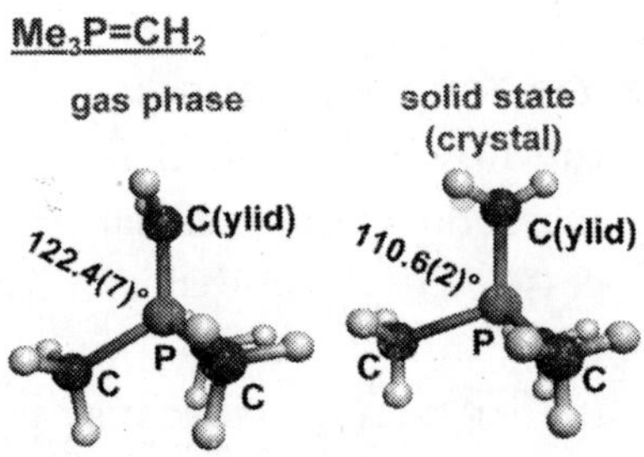

■ Z form

a duplex DNA structure in which there is the usual type of hydrogen bonding between the base pairs but in which the helix formed by the two polynucleotide chains is left-handed rather than right-handed.

■ Z-DNA

a left-handed variant of the DNA double helix originally observed in a sequence of alternating G and C bases, but possible in sequences of alternating purine and pyrimidine bases. The purines are 'flipped' 1808 about the glycosidic bond to assume the syn conformation (as opposed to B-form DNA, in which they assume the anti conformation); the pyrimidines remain in the anti conformation.

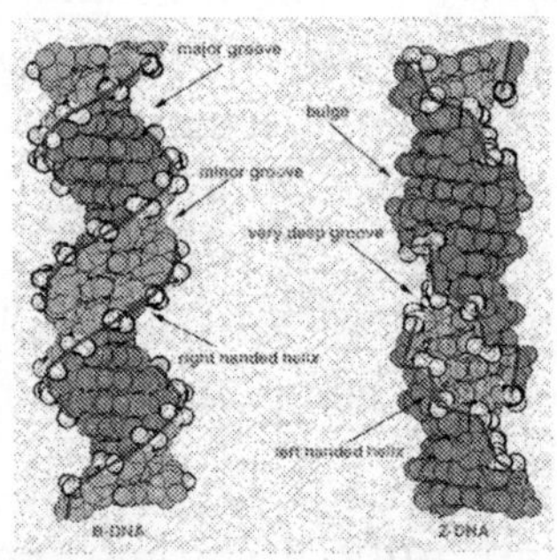

■ zebrafish

danio rerio. A model for genetic embryological analyses, which is unique among vertebrates in shortness of generation time, number of progeny and relative ease of examination of embryos.

■ **zero-order kinetics**

the insensitivity of a rate on the concentration of a reactant; especially for an enzymic reaction at substrate concentrations much above the Km.

■ **zinc finger (metal-binding finger)**

a motif of which seven or more may appear in a DNA-binding protein. Each is characterised by two closely spaced cysteine and two histidine residues that serve as ligands for a single Zn2+. When bound to Zn2+ they form a module from which protrude amino acid side chains that interact with the bases partially exposed to the DNA major groove. Since in the area of contact there is a limited and local parallel of DNA and protein sequences, the occurrence of the contacting amino acid residues in the protein sequence falls into a regularity, or register, that creates in the protein a syllabicity.

■ **zipper interaction**

the sequence of events by which an immunoglobulin G-coated lymphocyte draws the macrophage plasma membrane around it. Fc receptors on the membrane surface attach the immunoglobulin G markers and eventually surround the cell and engulf it in a phagosome.

■ **zona pellucida**

a glycoprotein coat that surrounds an oocyte or ovum.

■ **zonation**

the phenomenon by which specific areas of an organ have characteristic metabolic functions; e.g. alanine aminotransferase activity is relatively high in the cells of the liver served by the portal blood supply (the periportal region) whereas glutamate dehydrogenase activity is relatively high in the cells that feed the hepatic vein (the perivenous region), and glutamine hydrolysis and urea synthesis are prominent in the periportal region while glutamine synthesis is prominent in the perivenous region.

■ **zone of adhesion**

an area of a lipid bilayer where inner and outer leaflets

join, and by which proteins, gangliosides and other products may be targeted to the outer surface of the plasma membrane from their cytoplasmic sites of biosynthesis. Models postulate that such zones are continuous, e.g. they surround a pore through the membrane, or occur stepwise via a vesicle formed by the pinching off into the inter-leaflet space of part of the membrane inner leaflet, which can then fuse with the membrane outer leaflet.

■ zoo blotting

a technique to assess the degree to which a DNA sequence, usually a coding sequence, is conserved in evolution; a test for hybridisation of a probe complementary to the sequence in one species against DNA isolated from cells of a variety of other species.

■ Z-schaeme

in the Hill reaction of photosynthesis, the flow of electrons from water through photosystems II and I to reduce NAD+.

■ zwitterion

a dipolar ion with spatially-separated positive and negative charges. For example, most amino acids are zwitterions, having a positive charge on the α-amino group and a negative charge on the α-carboxyl group but no net charge on the overall molecule.

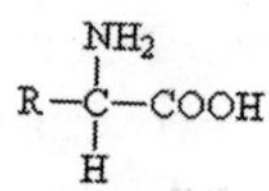

R is the functional group of the amino acid

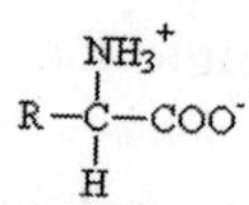

Ionization at physiological pH (7.4)

■ zygote

the product of fusion of a male and a female gamete; a fertilised ovum.

■ zymogen

a proenzyme; an inactive precursor of an enzyme, e.g. trypsiongen. Often, the zymogen is proteolytically activated by an endonuclease. In the case of some aspartate proteases, an acid-catalysed rearrangement allows cleavage and removal of an inhibitory C-terminal sequence, the prosegment.

■ zymogen granule

a cytoplasmic vesicle that contains a protein prior to secretion.

■ zymogram

a slab gel that, after electrophoresis, has been stained to reveal the presence of an enzymic activity, e.g. a gel infused or layered with a protein substrate and subsequently treated with a stain for the protein to reveal, as cleared areas, the location of proteolytic activity.

■ zymography

a histological method for localisation of an enzymic activity. The tissue or cell preparation overlays a gel that contains a chromophore- or fluorophore-labelled substrate, e.g. casein for a proteinase assay. After enzymic reaction, and washing or diffusing away of labelled products, the sample is coated with a photographic emulsion, exposed and developed to superimpose the histological sample on cleared areas in the labelled substrate that represent areas of enzymic activity.

■ zymosan

particles of yeast cell walls.

NOTES